The King's

GARDEN

The King's

GARDEN

Marguerite Duval

TRANSLATED BY

Annette Tomarken & Claudine Cowen

University Press of Virginia

Charlottesville

THE UNIVERSITY PRESS OF VIRGINIA
Translation Copyright © 1982 by the Rector and Visitors
of the University of Virginia

First published 1982

Originally published in 1977 as
La Planète des fleurs
by Marguerite Duval
© Éditions Robert Laffont, S.A., 1977

Library of Congress Cataloging in Publication Data

Duval, Marguerite, 1912-
    The King's Garden.

    Translation of: La planète des fleurs.
    Includes index.
    1. Botany—History. 2. Botany—France—History.
3. Botanists—France—History. 4. Muséum national
d'histoire naturelle (France)—History. I. Title.
QK15.D8813   580'.92'2   [B]   81-15934
ISBN 0-8139-0916-3   AACR2

Printed in the United States of America

CONTENTS

INTRODUCTION

THIS TRANSLATION of Marguerite Duval's *La Planète des Fleurs*, first published in Paris in 1977, is an important contribution to a sadly neglected facet of history—that of botanical exploration and discovery. During a quarter millennium of the age of discovery, a stream of daring and dedicated men searched the entire world from the Arctic Circle to the Antipodes to discover and bring back to Europe exotic and other useful productions of nature from every green corner of the world. No important aspect of history has been more unjustifiably neglected than the heroic, sociologically highly significant, often dramatic, often tragic, voyages of discovery of the plant hunters. Many of the treasures that were the fruit of their expeditions into the far-flung, newly discovered reaches of the world wrought immensely important changes and developments in the burgeoning Western civilization.

Although the trackless deserts and jungles penetrated by those intrepid explorers of the green world fought back less overtly than the red men of the Americas, the blacks of Africa, and the brown and yellow people of Asia and the Pacific islands, the resistance met by the plant seekers was at least as lethal; and, while the results of their discoveries lacked the drama of a Balboa proclaiming the discovery of the Pacific from the backbone of the Isthmus of Darien, or of a Pizarro plundering the Andean Inca Empire, the plant hunters were often of equally great significance in the development of Western civilization. Witness the sociological impact of the quinine-bearing cinchona on the control of the worldwide scourge of malaria, the demographic effects of the importation of the misnamed Irish potato on Western Europe, and the contributions of the sap of the rubber trees of the Amazon basin to the development of the machine age of the twentieth century, to mention but three among the many fruits of the plant hunters of a single continent, South America.

To document with specifics the quiet but profound effects of their contributions, one needs only a thoughtful visit to a produce market. I have just made a cursory inventory of the offerings of a nearby small establishment. Its bins, boxes, and display tables were virtually an exhibit of the utilitarian fruits of worldwide botanical discovery and exchange. There, dominated by members of the mustard family, were the Eurasian contributions of turnips, cabbages, cauliflower, mustard, radishes, and collards; lettuces and artichokes of the Aster family; three kindred items—parsley, carrots, and celery; and onions. With their historical roots in Asia Minor and the Mediterranean were apples, plums and pears, grapes and East Indian oranges, lemons and limes, most of the melons, including the watermelons, honeydews, and their cousins the cucumbers. There were tangerines with Philippine ancestry and eggplants with East Indian forebears, peaches with Persian ancestry; bananas and okra from the Old World tropics. The major New World offerings were Indian corn, tomatoes, peppers, Irish potatoes, sweet potatoes, and peanuts, all with South American origins. In the next bins there were Oriental chestnuts, English walnuts, Brazil nuts and peanuts of Brazilian ancestry, Mediterranean almonds and native North American pecans. Other North American offerings include avocados, scuppernong muscadine grapes, Jerusalem artichokes, and a variety of beans and squash.

A similar inventory of the shrubbery and flowers that embellish our gardens presents, just as overwhelmingly, exotic plant populations. Here in the South flamboyant oriental crape myrtles, azaleas, rhododendron, and wisteria dominate the landscaping. Only the shade trees are predominantly native American. Commonly, the most prized specimen plantings include a sweet olive or two, several camellias, and a mimosa, all three of oriental origin and all three first introduced into America by the French explorer-botanist André Michaux, whose life of botanical exploration is the subject of one of the chapters of this book.

During more than three exciting centuries, France and Frenchmen dominated the field of plant hunting and international botanical exchanges. Most of the voyages of botanical discovery were French-sponsored until Captain Cook's first voyage to the Pacific and the establishment of Kew Gardens, when England surged into rivalry for leadership in botanical discovery. The all but incredible accounts of those who played major roles in establishing and maintaining that long French dominance and the voyages they dared, such as those of Humboldt and Bonpland in South America and those of Jacquemont into Asia, as well as the careers and contributions of such great naturalist mentors as Buffon and the Jussieus, are told in these pages in lively style, contributing significantly toward remedying the persistent neglect of a major facet of the history of Western civilization.

HENRY SAVAGE, JR.

October 1981

The King's

GARDEN

HARDLY ANYTHING HERE GREW IN THE SOIL OF ANCIENT GAUL.
It all had to be found and brought back,
mostly by missionaries and botanists,
from Greece, the East, Persia, China, and the Americas.

[I]

The Montpellier Garden

HY was it not until the sixteenth century that the idea began to spread of going to find the world's flowers where they grew, in every land? The answer is relatively simple. Although there was much travel in the Middle Ages, those traveling rarely looked at the earth they trod. Rather, they moved across it, eyes raised heavenwards, anxious only to spread the good news of the word of Christ, blind to everything that was not immediately concerned with God. Such an attitude resulted in an abundance of martyrs and a shortage of observations. Here and there one finds a few scribbled notes, often indecipherable, usually having to do with edible plants, the only ones accorded any attention. The main texts continued to speak of Heaven and of Heaven's gradual victory over darkness. For these inspired writers the earth was a vale of tears. What those tears watered was of little importance.

The Crusades, however, did arouse some men's curiosity. These were holy wars, but trade was flourishing. Those who made profits through the combination of holiness, silk, and spices dreamed of going farther. It fell upon the missionaries to make the travel routes passable, pushing ever farther afield. Duly authorized by Rome, they ventured into the heart of Asia. Some managed to return to tell their tale. The boldest were undoubtedly the Franciscans and the Dominicans, who beheld sights unimaginable to Western Christian minds and encountered unbelievable beliefs, returning with terrifying legends and a certain amount of information.

It was this information, more than anything else, that attracted the attention of the popes, who sent more and more missionaries along the routes, first to China, which, because it had millions of souls, represented promising ground for evangelization. India also attracted them, and in 1328 a group of monks reached Lhasa, in Tibet. "The town is very beautiful," wrote one monk, "made entirely of white stone, with well-paved streets." Elsewhere they found a land "rich in corn, gold, and silk." They said nothing about the vegetation, but marveled at the diversity of trade. They discovered with amazement the great caravan routes, the complex marketing of fabrics, and medicinal drugs.

At the same time there occurred in France in 1349 an event destined to be the starting point of our story: Philip VI of Valois, king of France, bought the town of Montpellier from the king of Majorca. Even at this early date Montpellier had a fine university tradition. The first Faculty of Medicine in Europe was founded there in 1141, as were law schools. The remainder of the university dates from 1289. By 1349 the city was a cosmopolitan center, with all the cultures of the then known world. Young scholars there spoke Hebrew, Greek, Latin, and Arabic. Arab science predominated. In thirteenth-century Montpellier students were busy analyzing and testing; they distilled nitric, hydrochloric, and sulphuric acids under the guidance of a bold chemist named Arnaud de Villeneuve. In the fourteenth century, surgery was regularly performed and it was not unusual for corpses to be dissected clandestinely in order to further understanding of human anatomy. Another form of medicine based on plants was also practiced.

This kind of medicine derived primarily from the Arabs, past masters of the art. Constant reference was made to Avicenna, whose

Canon of Medicine had for some time been the basis of all medical study and who perpetuated traditions yet more ancient, in particular those of Egypt and Greece. Already there were botanists at the university studying "drugs," as pharmaceutical products were called at the time. The reputation of these men was so widespread that young people flocked from all over Europe to follow their courses, taught in Latin. But their work, based in the laboratory, was necessarily restricted to material provided by the immediate environment. Its direction was determined by a search for what might be useful.

Time passed, bringing the discovery of printing. After the appearance of the printed book, a number of authors of Greek antiquity finally became available to students and scholars, either in Latin translations or in French. By reading the classical authors, students of botany came to understand the full range of their subject. Although the Christian Middle Ages had been little interested in flora, the same had not been true of ancient Greece, which had produced numerous treatises on the topic. Such books were eagerly read by the young students of Montpellier, who devoured everything, chaotically, in whatever order the works happened to reach them, from Aristotle to Pliny, Hippocrates to Theophrastus, Dioscorides to Galen. If we wish to understand the mental point of departure of the travelers whose journeys we propose to follow, we must first pause to look at what they were reading at the time.

Hippocrates, the first of these authors to arouse enthusiasm, even eclipsed Avicenna. Born about 460 B.C. and living into his eighties, Hippocrates tried everything. Leaving Cos, his native island, he traveled throughout Egypt in search of the most ancient papyruses, shut away in the temples of the Nile. The Egyptians received him with respect, allowing him to consult all the papyruses he wished. He also learned much from the embalming sessions he attended. His readings taught him everything then knowable about the human body, including the role of the veins and arteries. In Egypt Hippocrates also found out the names of the plants employed to treat various organs, and the use of the poultice. A devout believer in medicine, Hippocrates felt that as a doctor he was exercising a sacred duty. Consequently, he was an extremely conscientious man, who could often be found seated beneath a plane tree meditating, reluctant to pronounce a hasty diagnosis. He had seen too many of his colleagues decide upon a diagnosis in terms of the plants at their disposal, allowing the remedies available to determine their assessment of the nature of the illness. Chance could thus dictate life or death. For this reason, Hippocrates devoted much of his time to studying plants, assisted by Crateuas, his herbalist. "You must pay careful attention," the master would warn. "I have discovered that plants which normally cure may kill if they have been contaminated—by snake poison, for example. This poison can kill in a decoction of leaves or flowers." Hippocrates himself mentioned only 234 plants, but Crateuas probably listed many more in texts, which have not survived. Still extant are Hippocrates' observations on wormwood (*Artemisia absinthium*), on the seeds of the castor plant (*Ricinus communis*), and the pomegranate (*Punica*), on vine leaves (*Vitis*), the mandrake (*Mandragora*), and what he called the hellebore.

Theophrastus, c.372–c.287 B.C., was more a philosopher of nature, insofar as the distinction between scholar and philosopher had any meaning at that time. Few authentic texts by him survive, his work having attracted too many commentators and translators. What we do know about him is that this universal scholar, a disciple first of Plato, then of Aristotle, sought to situate plants in the overall context of the universe. A statement by Aristotle particularly disturbed him. "We find in plants," the philosopher said, "a continuous ascent toward animal life." Theophrastus was probably the first Greek scholar to plant his own botanical garden.

Still, all this transmitted knowledge was somewhat vague. Different was the case of Pliny the Elder (23–79), the Latin author whose *Natural History* brings together the knowledge of his day.

"I offer," he wrote, "more than twenty thousand items of information, in thirty-six volumes, the result of my reading of more than two thousand works by a hundred carefully chosen authors, few of which are accessible to the curious reader, because of the difficulty of the subject. To these sources I add much information unknown to our ancestors. . . . And I am sure that I have neglected a number of other writers. Busy with my official duties, I write my book in my spare time, that is to say, at night." In a way, Pliny had composed the first encyclopedia. So far as plants were concerned, Pliny considered them from the varying points of view of philosophy, history, medicine, magic, and agronomy.

But the man to whom the first Montpellier botanists owed everything was Dioscorides, their constant reference. He was the true father of the future science, but we know little about him. A doctor, he lived in the first century A.D. and left behind him a treatise entitled *De Materia Medica*. These are the only certain facts about Dioscorides. But numerous legends concerning him circulated. It was said, for example, that he had lived at Cleopatra's court in Anthony's day and that a mysterious Byzantine princess was the guardian of his principal manuscripts. His highly individual handwriting did make possible a collection of the texts composing his treatise, which was published for the first time in Venice by Aldus Manutius in 1499. A French version by Jean Ruel appeared in France in 1516. Everything began with this edition. What did Dioscorides say to arouse such enthusiasm?

First, he revived in a remarkable manner the spirit and inspiration of ancient, pre-Roman Greece. Probably he contributed to the renewal of Hellenism in the first and second centuries A.D., a movement that facilitated the link with modern times. About plants, he put forward an idea that was to astonish the botanists of Montpellier. He affirmed that if one wanted to understand the life of plants, one had to observe them in situ, patiently and at length, without cutting them. When dead, they could no longer teach us anything. Nowadays we might understandably be surprised at the amazement caused by such a statement. Botanists had just not thought of such a possibility. Busy with their pharmaceutical equations, they gathered more and more plants (or, rather, had others gather plants for them), and discoursed endlessly about dried flowers.

Dioscorides, on the other hand, examined growing plants. He would spend many hours sitting beside a specimen to observe its changes and formulate detailed descriptions. He also studied fruits on the trees, the same fruit for several days, and then the seeds hidden deep inside the fruit. He discussed in this manner pomegranate (*Punica granatum*), medlar (*Mespilus germanica*), and lemon seeds. He examined roots of the milkvetch (*Astragalus*), the laurel tree (*Laurus nobilis*), ginger (*Zingiber officinale*), gladiolus (*Gladiolus byzantinus*), the hyacinth (*Hyacinthus romanus*), the narcissus and water lily (*Nymphaea alba*). He described the scent of flowers, their shades of color, and their beauty. The beauty of flowers was certainly something scholars had never considered. This idea was to open new horizons and make men long to move outdoors.

But Dioscorides had already warned that beauty should not be equated with innocence. The new domain had a sinister aspect. Certain plants could kill. They even destroyed one another. Scorpions and tarantulas were not the only producers of poison. Dioscorides succeeded in brewing poisoned honeys and appropriate antidotes. These were dangerous games, uneasy pleasures, but this is how knowledge is acquired. Dioscorides traveled. He went to Arabia, where he conducted a curious experiment on trees of the turpentine family (*Pistacia terebinthus*). Cutting grooves into the trunks, he collected the resin, which, when heated to 100°C, gave off the mysterious perfume of incense. In Egypt he drew up a list of oils made in that country; in Greece he studied olive trees. He investigated salves, which he concocted with ingredients of all kinds, from quinces to lemons, from wild vine leaves to Persian jasmine

(*Jasminum officinale*) and myrrh (*Commiphora abyssinica*). He discovered the composition of vitriol and examined mosses. The list could be lengthened considerably.

"Dioscorides," wrote Galen, "condensed into five volumes all the useful material relating not merely to herbs but also to trees, fruits, flowers, juices, and syrups. In every case, he seems to me to succeed better than anyone else in describing the composition of remedies." This homage from a contemporary is all the more precious because Galen, a great scholar, was to exert a profound influence on fourteenth-century France.

Born in Pergamum, Galen (130–200) studied philosophy and medicine. By performing numerous dissections, he was able to make important discoveries in anatomy. Since, at the time, knowledge was unified, he was also interested in arithmetic, geometry, astronomy, and botany, for which he had a particular fondness. He read much, but felt it important to test the truth of his knowledge. Accordingly, he decided to travel in order to see plants growing in their natural environment. He went first to Palestine, visiting the mines of Cyprus on the way. Then he journeyed through Cilicia, Phoenicia, and Crete. When he returned to Rome, he was able to describe the properties of more than four hundred plants. His study was enriched by observations on the warmth, cold, or humidity of the atmosphere in which the species grew, factors which, he declared, considerably modified the plants' medicinal properties.

This work was yet another invitation to travel in order to analyze something one had not first cut off from life: "I do not believe," Galen declared, "that accurate descriptions can be made except in front of the living shrubs, flowers, and trees." He named a number of plants, the autumn crocus, for example (*Colchicum autumnale*), which grew in Colchis, the land of Medea, the enchantress, and contains a dangerous poison. Galen also gave names to the euphorbia and the gentian. The plant he described could be *Gentiana pannonica*, *G. phlogifolia*, *G. lutea*, or *G. purpurea*. He

7

also listed a flower which he called heliotrope, a name later given to a species native to Peru.

It is important to understand the effect such writing and practices might have had on young scholars who rarely ventured out of their laboratories. In lively polemics the upholders of the dissection tradition opposed those who heard this call from ancient Greece inviting them to take to the open road and to abandon plants dried in the laboratory in favor of the more tantalizing possibilities of work in the field. The time was ripe for such ventures. In the sixteenth century the world was opening up. And so the traveling botanists, as they were called in Montpellier, set sail.

Montpellier, the first botanical garden. Etching attributed to Richer de Belleval. (Montpellier, Jardin des Plantes. Photo by ERL)

Pierre Belon. (Paris, Bibliothèque Nationale. Photo by ERL)

[2]

Pierre Belon,

First Botanist-Traveler

Pierre Belon, destined to become the first official botanist in French history, was born in the hamlet of La Soultière, in the province of Maine, in 1517. We know little about his childhood except that he acquired a thorough knowledge of the fauna and flora of Brittany, where he spent much of his adolescence and whose language he spoke. Later, at about the age of twenty, he went to the Auvergne to work as an assistant to an apothecary attached to the bishopric of Clermont. He concentrated there on zoology, studying the presence of the "medium brown vulture" on the Mont-Dore mountains and the trout in the Clermont River. He never completely abandoned this interest and published in 1553 a work entitled *De aqualitibus* and, in 1555, the *Histoire des Oyseaux*.

After 1535 René du Bellay, bishop of Le Mans and a great lover of botany, took Belon under his protection. The bishop's palace was in Touvoie, and it was doubtless there, among the flowerbeds, that botany finally won Belon's heart. On du Bellay's advice, he left Touvoie to go to Germany to Wittenberg University, where Valerius Cordus offered a popular course in which he undertook a

9

"demonstration and interpretation of the plants of Galen, Theophrastus, and Dioscorides."

Wittenberg was then the stronghold of Lutheranism and on several occasions Belon was able to discuss religious matters with Luther himself. But most of his time was occupied with the teachings of Cordus. The master and pupil having formed a friendship, Belon soon became a traveling companion to Cordus, who had a taste for nomadic life. We find them first gathering plants "for four months in all the states of Germany." After this trip, Belon returned for a rest at the home of his protector, du Bellay, who decided to finance his studies in Paris, where he spent the year 1542. A year or two later Belon left to rejoin Cordus, who, with some assistants, was plant hunting in Italy. But Cordus died in Rome in September 1544, leaving Belon to retrace his master's itinerary alone. During this journey he visited Bembo in Padua and the Venice Botanical Gardens, where Messer Aloisio, the gardener, introduced him to acclimatization techniques.

When he returned to Paris he entered the service of the cardinal de Tournon. Now with access to the court of the Valois he could develop at leisure in that brilliant, brutal, and sophisticated society. Francis I, whom Belon soon approached, made a profound impression on him. Interested in everything, the king appeared to the botanist a "matchless conqueror of all animated substance." Was it not said that the monarch sometimes amused himself by having a real lion sleep on his bed, and that the lion would then purr like a common cat? Moreover, had the same king not gathered together in a special room all the "curiosities" presented to him by scholars, travelers, and diplomats? Francis had his orangeries, his menageries, and his heronries. And there were his gardens at Fontainebleau, adorned with rare flowers, some even coming from the newly explored Canada.

Nevertheless, Belon grew bored. After much time spent pacing the forest of Fontainebleau, he decided that sedentary research was not to his taste. He needed drama and adventure, the heady excitement of discovery. He dreamed of Greece and of plants which, in his day, were only magic-sounding names in Dioscorides' classifications.

An opportunity for change soon presented itself. In the East, the king's affairs were going badly, and Francis was anxious to keep the friendship of the Grand Turk, Soliman the Magnificent. In 1546, he decided to send to Constantinople a large embassy led by M. d'Aramont. In order that the group might appear more important, every effort was made to give it a "scientific and literary" dimension. Thanks to the cardinal de Tournon, Belon was appointed a member of the group of envoys. In a letter to his protector he made clear his pleasure at the appointment: "After you learned of my desire to understand matters relating to the nature of medications and plants (an understanding I could not really acquire from anything but extensive travel), you deigned to command me to go to see these plants in distant regions, seeking them in their places of origin. I would have been unable or hesitant to undertake this journey without your help."

The mission set off at the end of December 1546. On February 9, 1547, they reached Venice, where d'Aramont equipped three galleys for the journey to Turkey. They put in at Ragusa, where Belon found the horsetail (*Ephedra distachya*), an extremely ancient plant, probably originating in China and one having powerful medicinal properties. The plant delighted Belon, who thoroughly explored the coast and nearby islands, visiting temples and monasteries. He had not lost his interest in fish. On March 13 d'Aramont continued his journey by land. Belon, less restricted by diplomatic duties, chose to visit Greece and then rejoin the mission in Constantinople.

His next stops were Corfu, Zante, and Kithira, or ancient Cythera. Near Corfu, Barbary pirates carried off Belon's companions, leaving him in an empty boat. He reached Crete, then under

Venetian rule, and was most courteously received. The nobleman Antonio Calergo of Candia supplied him with guides for the ascent of Mount Ida, where he thought he would be able to gaze upon "Jupiter's tomb," but he met mainly bats. In Crete he also witnessed the harvesting of labdanum (*Cistus ladaniferus*), "one of the most famous drugs among our perfumes." Labdanum, not to be confused with laudanum, was at that time a specialty of Crete. An odorous resin, it produced an oil and probably a wax. Suitable for cleaning, strengthening, and dissolving, it was used as a poultice and, most importantly, as a fixative. Groups of traffickers would steal it to sell at exorbitant prices to perfume merchants. Labdanum was considered so precious that anyone caught stealing it was condemned to death. After a long stay, Belon finally set off for Constantinople. At the end of April he was on the banks of the Euxine, at the mouth of the Bosporus.

In Turkey he frequented all sorts of secret shops, always carrying with him Avicenna's *Canon of Medicine*, which gave the Arabic names for combinations of plant extracts. This enabled Belon to formulate for himself a sort of glossary of the Turkish language. But the owners of these small shops, who were called "druggists," never gave away the secrets of their salves and ointments. If need be, they preferred to eliminate the indiscreet inquirer—indeed, a veritable network of spies peopled those narrow streets.

This inquiry was of great importance to Belon because France did not have direct commercial access to the medicinal drugs of the Levant. He managed to obtain several pieces of Lemnos earth (*Lemnia sphragis*), said to cure the plague, and even visited its source on Lemnos. From there he at last reached Mount Athos, which he found to be a "paradise of delights for those liking to be out of doors." He marveled at finding, on the mountain oaks, a mistletoe "completely different from the one we see growing on apple trees" in France. In the same place grew the white helle-bore, or Christmas rose (*Helleborus niger*).

In Salonika he was moved to see again the navelwort (*Cotyle-don*), a common plant in his native Maine. Then he visited Kavalla, the major port of Macedonia. At the beginning of August 1547 Belon returned to Constantinople, to set off again in the company of M. de Fumel, another diplomat, and his attendants. The group was heading for Egypt, but they stopped along the way to visit the Greek Isles. On Chios, Belon examined plantations of mastic trees (*Pistacia lentiscus*) and he gathered a little of the "green earth" of which the island contained such a precious deposit. On Cos he evoked the memory of Hippocrates, born on the island, and fell in love with the plane tree, which he dreamed of constantly from then on. After Rhodes, he reached Egypt, where he landed at Alexandria. There he was received by Benoît Badiolus, consul of France and Florence. Then, still with his companions, he traveled toward the mouth of the Nile, where he was shown, as curiosities, the branching palms (*Hyphaene thebaica*), "some of which," he noted, "bear on a simple trunk twenty large trees, each separated from the other but all having the same origin at the foot of a stump." At Rosetta he picked "papyrus grass," the reed that gave paper to the Egyptians. Crocodiles also fascinated him, those "peculiar offspring of the Nile." He observed the cultivation of taro (*Colocasia*), an edible plant; of the banana tree (*Musa sapientum*); of sugar cane (*Saccharum aegyptiacum*), which came from India; and of the *Ficus sycomorus*, with its "exquisite greenery." He loved Egypt.

After a brief visit to Cyrenaica, he returned to Cairo to join de Fumel, who was preparing to leave for the Sinai. On the edge of the Red Sea, Belon dug up some roses of Jericho (*Anastatica hierochuntica*) and picked what was probably "bastard senna" (*coronilla*). The expedition lasted twenty days. The travelers next decided to go to Jerusalem, but first they had to confront the desert, the caravans, and the bandits. It was, wrote Belon, "a strange and difficult journey." In Jerusalem he inquired about henna (*Lawsonia inermis*) and searched the thorn thickets "to find out what species of thorn was used to make Christ's crown." Then came the Holy Sepulchre and the river Jordan. More robbers, caravans, and deserts. They passed through Bethlehem and, moving along the banks of Lake Tiberias (the Sea of Galilee), they took the road to Syria and arrived in front of the iron-clad gates of Damascus six days later. The Syrian orchards filled Belon with admiration. The vines there were magnificently cultivated. He was dazzled by the plentiful pear, apple, apricot, and almond trees. He noted everything, even the fact that the Syrians grew fig trees in groves.

They set off again, this time in the direction of Lebanon. Robbers once again attacked the group, but they reached Aleppo, where Belon was fascinated by the trading—or trafficking—in the various juices, gums, and resins that supplied the druggists of Europe. Then they entered Cilicia, where Belon daydreamed, sitting beneath plane trees, cedars, and cork oaks. Finally they returned to Turkey and since it was winter now, he could rest a little.

Belon did not return empty handed: From his wanderings he brought back a large crate filled with snakeskins, "birds, land animals, entire plants, seeds of unusual plants, and several marine creatures." He entrusted the crate to a Genoese vessel, the *Delphina,* which set sail for England, but pirates attacked and "We lost everything," noted Belon sadly.

At last he returned to France and the court. Francis I had died during Belon's absence, and Henry II was now king. From the Orient, Belon had, in the end, brought back only a few seeds he had managed to save and some sketches. He had other compensations, however. He was brimming over with images, ideas, and stories. Because he had seen them himself, he could describe marjoram (*Origanum*); gentian (*Gentiana*), used against the plague; mallow (*Malva*); and hyssop (*Hyssopus officinalis*), the Hebrews' sacred herb, placed by King Solomon in his bath and supposed to cure colds and bronchial infections. Belon could recall the reddish color of henna, a shade familiar to us; the beautiful myrrh, whose resinous juice cured pains in the joints; and the small poppy, which alleviated pleurisy. He was haunted by the white hellebore, but he knew that one must beware, because the Christmas rose is poisonous. It was reputed to drive men mad, although others said that the plant cured madness. In general, the more beautiful the plant, the more formidable it was. The celandine, for instance (*Chelidonium majus*), was a small, yellow-flowered plant upon which one could not walk barefoot without risking death. In Egypt the loveliest peonies (*Paeonia*) were used against coughs, and Belon was told that the local thistles (*Cnicus benedictus*) were meteorologically sensitive, foretelling sunshine or rain by opening or closing their flowers. Other thistles, apparently, could burn a shepherd's crook. Then there was the mandrake (*Mandragora autumnalis*), against which he had been earnestly warned by the monks at one of the monasteries he visited. The plant played an important role in medieval sorcery. Was it male or female, men wondered, and what did its roots do? People were not sure, but they remembered old beliefs. Pliny, for example, recommended that one never touch a mandrake without first drawing three circles around it with the point of a sword.

Another adventure remained for Belon. This undertaking ratified and, to a certain extent, justified the entire botanical enterprise: It concerned the acclimatization in France of seeds collected abroad. In Belon's day, this art was still in its infancy. For the most part he failed in the endeavor, though through no fault of his own. He offered Henry II the objects he had brought back from Asia and explained his plans to the monarch. He proposed to acclimatize —to "tame," as he put it—on French soil, several useful and pleasant plants, such as the holm oak (*Quercus ilex*), the cork oak (*Quercus suber*), the spruce (*Picea*), the strawberry tree (*Arbutus unedo*), the turpentine tree (*Pistacia terebinthus*), and the Sicilian sumac (*Rhus coriaria*). The king was enthusiastic and he awarded Belon an annual pension of 600 livres in a royal warrant. But making a promise does not always insure keeping it. As ever, the king's coffers were empty. While he waited, Belon gathered plants, but from less far away. He went to the Auvergne to make an inventory of forest flora and then to Switzerland, where, in the spring of 1557, he marveled at "the trees rejoicing in the beauty of their flowers and the birds singing loud in the burgeoning forests." In 1558 he traveled again in Italy, primarily in Tuscany, where he visited a number of gardens to continue his study of the problems posed by acclimatization. He was presented with two fine cuttings of cherry laurel (*Prunus laurocerasus*), but he had to cross most of Italy, "reddened with the flames of war," to find, near Sienna, acorns of the cork oak.

Despite many supplications and protests, the royal award never came. Belon, who was not without ambition and who had formulated a number of concrete, sometimes bold proposals, such as the introduction into France of poppy culture, which would have made possible the use of opium for pharmaceutical purposes, had to content himself with acclimatizing what, when, and where he could. It was to be at Touvoie, in the house of his first protector, René du Bellay, now dead, that Belon was to work most frequently. There he sought in particular to "tame" his beloved plane tree, whose growth he watched anxiously. "And so, Platanus, you who chose for your wild habitat the valleys of Asia, a colder climate than ours, why should we not be able to have you here? We think, therefore, that you will not be so obstinate as to prevent our enjoying you without having to spend a thousandth of the money paid by a single Roman citizen to bring you across the sea. For, since you have grown from seed up to the fifth leaf, we hope that you will not slip away from us in the winter. And if we once have a dozen of you, we will one day have a thousand."

Thanks to Belon, winters at Touvoie were brightened by the presence of the Christmas rose, brought back from the Bergamo area. Many of Belon's trial plantings have not survived because in his day there were no special gardens adapted for the care and cultivation of plants and seeds. Some traces of his activity do remain, however, most notably in Paris. To seeds he brought back from Italy we owe the naturalization, in several gardens of the capital, of the manna tree (*Fraxinus ornus*), or flowering ash; the sumac (*Rhus coriaria*); the olive tree (*Olea europaea*); the plane tree (*Platanus orientalis*), of course; and the oleander (*Nerium oleander*). His attempts met with varying degrees of success but nonetheless justify our placing Belon firmly among the fathers of arboriculture, just as he was the first of the great French botanical travelers. On both these counts, Pierre Belon was a true pioneer, and the masters of the future science have never forgotten their debt to him. We must, however, consider both the positive and the negative aspects of his career.

Belon lived at one of the major turning points of history. Still imprisoned by the vague approximations of ancient science, he had no other guidelines than those supplied by his distant predecessors, whose heritage, for the most part, was based on allegory and legend. In Belon's time, for example, people could not even distinguish a male from a female plant, for nothing, or almost nothing, was known about plants' reproductive methods. Belon also believed that every plant name should be "rendered in its majesty by way of antiquity." He lived in this world of legend: In the last analysis, he scarcely went further than Dioscorides. Although he does mention 275 Levantine plants, he does not really describe them, preferring to adopt the classical nomenclature. He has also been reproached with the fact that, on his return from the East, he proved himself more practical than experimental. He was accused of being interested mainly in the therapeutic and nutritional uses of plants, in techniques for cultivation and yields and costs, and so neglected pure science. This criticism ignores the fact that these practical preoccupations were one of the reasons for Belon's mission. France had been sorely tried. It was in urgent need of everything.

To Belon's credit must be placed his general approach. A man of adventure and of knowledge, he loved plants as he loved birds and fishes. He went to see them where they lived, and he described the existence of plants and flowers the West had but dreamed of. He had seen for himself the roses of Jericho, the larkspurs on Mount Ida (*Delphinium ajacis*), hellebore, acacias, myrtles, rock roses (*Cistus*), and mighty cedars. He would undoubtedly have set off again in search of the plants he had lost at sea, if there had been time. But at the age of forty-seven, in 1564, he was assassinated in the Bois de Boulogne, near the Château de Madrid, where he had been installed by Henry II. His death was attributed to a passing prowler. Because the police in those days did not pursue investigations, the matter rested there. But Belon's friends were convinced that his death was connected with his travels. He knew too much, they thought, about the rarest plants, their medicinal powers, their possibilities for adaptation, and, above all, about the manner in which they were traded.

The medicinal plants section of the King's Garden in 1636.
(Paris, Bibliothèque Nationale. Photo by ERL)

The West Indies paradise. Drawing by Father du Tertre.
(Paris, Bibliothèque Nationale)

[3]

The Creation of the King's Garden

ANY were to follow in Pierre Belon's footsteps. But before describing their quest for plants, let us pause to consider briefly the state of botany in the sixteenth century. It was then still a science-to-be.

In Germany, Otto Brunfels was born in Mainz at the end of the fifteenth century (he was to die in 1534). A cooper's son, he began to study medicine and soon discovered plants by way of the Greek authors, whom he translated into German. He put forward an opinion that was to give rise to endless discussion. He rejected the medicinal use of oriental plants, asserting that only indigenous plants grown in the soil of a given country could satisfactorily treat the inhabitants of the land. Brunfels, like so many of his predecessors, based his arguments upon Dioscorides. But the problem he posed provoked much reflection among philosophers: Can the men of the North, they asked, make wide use of plants coming from elsewhere; can they withstand the effects of medications destined for men whose organism, given variations in climate, vegetation, and tradition, might be very different from theirs? It was a delicate question, one not entirely answered in our own time. What medicine is good for whom?

During the same period another great German scholar, Leonhard Fuchs, a doctor of medicine and a fellow of the Academy of Tübingen, reflected upon the future: "I did not wish," he wrote, "that our descendants should suffer from the same handicap as we and the authors of Antiquity so obviously have done, namely, that species familiar to everyone today may rapidly become obscure for our successors." He undertook an extremely detailed description of all the flora of his country. He was the first to work closely with an accurate draftsman. The lesson was not lost—the French botanist Charles Plumier later gave the name of this pioneer to a flowering shrub, the "fuchsia," which comes from New Zealand and South America.

The Spaniard Andres de Laguna, physician to Pope Julius II, translated the Greek authors into his own tongue in the course of a journey to Rome in 1592. He claimed that, thanks to the pope, he had access to authentic manuscripts, including a treatise by Dioscorides on poisons and antidotes. The latter work created a great stir in Spain, where knowledge of medicinal plants was more profound than in other European countries because of the Spanish inheritance of Arab learning and its Mediterranean vegetation. In addition, Spain had just received from Christopher Columbus plants that set the world dreaming. The treatise on poisons, attributed to Dioscorides, impressed the Spaniards sufficiently for them to circulate it among their scholars and communicate its contents to all their navigators.

In England a man of exceptionally wide learning, William Turner, a doctor of medicine but also a philosopher, art collector, and owner of a distinguished library, decided that he should help increase the knowledge of plants in his country. After reading Fuchs and using his plates, Turner began, in 1551, to regroup existing knowledge by classifying plants in an alphabetical order. Furthermore, he declared war on Latin plant names. He believed them detrimental to the spread of botany because they necessitated research he deemed "shocking and inhuman." Accordingly, he began by setting up an index of names in Latin, but, to facilitate the identification of plants, he translated these names into Greek, English, German, and French. This was a decisive step toward the production of modern books and works of popularization.

Meantime in the Low Countries lived a doctor and enthusiastic botanist, Rembert Dodoens. Born in Mechelen in 1517, Dodoens studied at the academies of Germany, France, and Italy. Invited to the Academy of Leyden, he settled to teach and wrote books on the medicine and plants known in his day. He was particularly interested in those European plants whose pharmaceutical properties were currently being discussed, and in flowers and cereals. Because of his work, government authorities began to cultivate useful plants. One day, a variety of tulip freshly arrived from Turkey caught Dodoens's attention. He recommended its acclimatization. We all know the sequel to this incident. Dodoens, although consulted by sovereigns, worked in Leyden until his death in 1585.

In Switzerland, Konrad von Gesner, a botanist's son from Zurich, was the first scholar of botany for internal use. Particularly fond of mountain plants and flowers, he traveled all over the Alps, where he discovered more than five hundred hitherto unknown species. He would plant these in his garden to observe how they grew out of the ground, budded, opened, and bloomed. He did something no one had done before him: He tried on himself all the medicinal plants he found. His notations always began with the taste of the plants. He died of the plague in 1565 at the age of forty-nine.

In Italy the art and science of botany were the most highly developed, chiefly because there had been no break there between antiquity and the modern era. All the Roman emperors had been interested in medicinal plants and had sent for them from where they grew. In addition, numerous ancient manuscripts had been brought to Rome when Byzantium fell. And gardening had been an art in Italy long before the rest of Europe dreamed of taking an interest in such matters. Italy was glutted with plants from every corner of the former empire, plants the very existence of which was unknown elsewhere in the West. The Italians of the Renaissance, however, were not concerned with the increasing interest in botany throughout the Western world or the quest for plants and their classification. They were preoccupied by the creation of the first true botanical gardens, which rapidly became legendary and were to arouse enthusiasm all over Europe.

In France, ever since the twelfth century, as we have seen, the old university city of Montpellier had boasted a faculty of medicine with a well-established tradition of independence. This faculty had resisted the passions of the wars of religion and, by the middle of the sixteenth century, constituted a permanent forum for bold and brilliant research and work. In 1550, during the reign of Henry II, the teaching of botany became an official part of the university's functions. A royal edict stipulated that "one of the most suitable and capable doctors of the university should be named to lecture to the students and place before them, from Easter to Saint Luke's day, herbs gathered in the said town of Montpellier and the surrounding districts."

A famous line of botanists was to be trained at Montpellier, beginning with Guillaume Rondelet, to whom the new chair was awarded. Born in 1507, in Montpellier, Guillaume Rondelet was first a schoolmaster. Later he moved to Paris, where he learned Greek. He conceived a passion for botany, earned his doctorate of medicine in 1537, became *professeur royal* in 1545 and chancellor of the university of Montpellier in 1556. The plants of the Greek islands and the Nile delta were relatively well known, but students knew nothing, or almost nothing, about what grew in France. Dr. Rondelet began to perform revolutionary work merely by taking his pupils out into the countryside. Before long one could see solemn, thoughtful young men roaming the fields and woods, gathering with infinite care flowers to which no one before had ever paid the least attention. A native of the area, he knew its riches and believed that the region surrounding Montpellier should, by virtue of its exceptional geographical situation, furnish a collection of plants which would represent a wide range of mountain, valley, and coastal flora. Beneath his affable exterior, Rondelet was a passionate and private man. He was convinced that it was, or would one day, be absolutely necessary to look at the inside of plants without destroying them. He urged his students to perform meticulous research. Eager to reach conclusive results, he required of his disciples not merely a written description but precise drawings of every part of any plant under consideration. He had no intention of seeing the work he had begun disappear upon his death, and so he observed his pupils with the closest attention in the hope that he might find one or more to whom he could hand the torch.

The heir of a noble family from the Armentières area, Charles de l'Ecluse, soon attracted Rondelet's attention. Born in 1526, de l'Ecluse obtained his bachelor degree in law to please his family. He knew Latin as well as the princes of the Church since he had learned it to fulfill the wishes of the grand prior of the abbey of Saint-Vaast, his uncle. He chose the Reformation, however, after studying at the university of Wittenberg, where he became friendly with Luther. In addition to French and Latin, de l'Ecluse spoke fluent Greek, German, Spanish, and Italian. It was the European reputation of its faculty that had attracted him to the universiy of Montpellier.

The man stood out from his fellow students. He was reserved, but he burned with an inner fire tempered by philosophy and reflection. His politeness and education made him approachable to all. His background was impressive, his knowledge of six languages dazzling. Last, he was available, not yet having decided upon a career, except for the fact that only concrete, detailed work could hold his interest. He certainly had not envisaged being a botanist or even a doctor. But when he discovered Dr. Rondelet's activities, he was immediately enthusiastic. Plants seemed to him to form a new realm midway between that of God and that of man—living, flowering, fading, and being reborn, but always defying man.

Guillaume Rondelet was soon captivated by this strange young man. He offered him lavish hospitality, and he made de l'Ecluse his secretary in order that the student might be able to follow all the work of the faculty. In his research Charles de l'Ecluse used and enriched the descriptive system recommended by Rondelet, but he went further. He invented a new descriptive terminology, and he detected, in certain plants, new parts, hitherto assimilated with large units. He described these as distinct, thereby making a significant step forward in the analysis and morphology of plants. Seeds interested him, as did sowing dates. Many plants died in those days because they were not sown at the right moment. Finally, providing the first important manifestation of continuity in botanical matters, he picked up again Pierre Belon's studies of the gums and shrubs that produce juices and resins, and he recommended the acclimatization of such plants. He was also a lover of ferns. He had friends everywhere, and he asked them to send him from far away everything he thought he might be able to grow in France. He attempted successfully, for instance, the acclimatization of the horse chestnut tree (*Aesculus hippocastanum*) which grew in Greece and the Near East.

But it is with the potato that his name will forever be associated. De l'Ecluse was, in fact, the first to grow it in Europe. From his English friend Sherard he received a sample of the variety "Papas," which had been brought back from Peru by British sailors. De l'Ecluse conducted some trials, tasted the new tuber, and sent it to his acquaintances in the major European countries. The Germans were the first to cultivate it, but only cautiously at the beginning. The potato had little success in France. It was not until the eighteenth century that Parmentier succeeded in introducing wide-scale potato cultivation. In the interim, other varieties had reached Europe. But Charles de l'Ecluse had some consolation: Word went around, and people spoke about "his" red potato, and it was planted in a few vegetable gardens.

In Rondelet's time the Montpellier School produced other important botanists, including the brothers Jean and Gaspard Bauhin, who together studied almost four thousand plants. This time the progress made was considerable. Beginning with the work of Charles de l'Ecluse, who had refused to study ancient documents, the Bauhin brothers reviewed, for every plant, and in a predetermined order, each of the external parts—root, stem, flower, fruit, and seed. This significantly refined the mass of observations previously accumulated. The process of ordering, carried out for the most part by Gaspard Bauhin, earned him the reputation of being botany's "lawgiver." He did indeed formulate the first laws, deciding, ordering, and distributing family names under which he grouped related species.

Other unusual persons frequented the corridors of the Montpellier faculty "and the surrounding areas." Two of these, Leonard Rauwolf and Pierre Richer de Belleval, were to have a deep influence upon the future of botany and botanists.

Although the wars of religion were devastating France, a liberal spirit still reigned in the faculty, where Protestant students remained free to work as they wished. One of these was the turbulent young German Leonard Rauwolf. Elegant, restless, and curious, Rauwolf hated to stay still. He longed for fresh horizons. Pierre Belon was a hero to him because he had traveled in the Levant. Rauwolf believed that it was time to go back "over there." He paid homage to Guillaume Rondelet, whom he called his "preceptor," but he added that because of Rondelet's work, much more was now known and a return to the Levant was essential. This idea became the new topic of discussion. Was travel or laboratory work the more fruitful, the scientists wondered, and was adventure not richer than experimentation? Rauwolf had no doubt as to the answer, but he insisted on the need to prepare future travelers thoroughly. They would have to be new men with the usual solid scientific background, but also conditioned psychologically to be able to face the many perils they would be exposed to and to defend themselves. They would have to possess a good knowledge of every country to be visited, its history, mores, and religion. Rauwolf was farsighted, and events were to prove him right. Botany would claim many martyrs, often because travelers were inadequately prepared.

Guillaume Rondelet, however, was too committed to conducting a census and analysis of the plants around him to yield to Rauwolf's arguments. The inventory of the flora was far from fully known. Why dissipate one's energies? And so, in 1560, the gifted student and traditional master clashed over this problem, destined to be an eternal source of debate in botany. For the time being, to please his revered "preceptor," Rauwolf agreed to collaborate on drafting a list of the flora of the Lyons region and to complete that of the area around Montpellier. When Rondelet died in 1566, he could be well satisfied with his work.

In 1573 Rauwolf could at last undertake his journey, one that would keep him far from Europe for several years. On his return, he published his *Itinéraire en Syrie, en Judée, en Arabie, en Mésopotamie, en Babylonie, en Assyrie et en Arménie*, a work that met with universal acclaim. He had collected many plants and had also done geographical and sociological work. In Arabia he had been fascinated by a strange little seed from which could be brewed a delicious beverage. This brew seemed to him to have an effect upon the mind, making it "curiously animated." The bean in question was *Coffea arabica*. This interesting discovery is discussed at length by Rauwolf: "Among other good things," he wrote, "the Arabs possess a beverage which they greatly prize, calling it '*chaubé*.' The liquid, black as ink, is extremely useful in treating various illnesses, particularly those of the stomach. The Arabs are accustomed to drinking it in the morning, in public, without fear of being seen. They serve it in small, very deep earthenware cups, drinking it as hot as they can bear. Often they lift the cup to their lips, swallowing the *chaubé* in small sips. They make the drink with water and the fruit the inhabitants call 'bunnu,' which resembles, in size and color, the berries of the laurel." Rauwolf was amazed that Pierre Belon had not already mentioned the custom. The Arabs were well aware that in the *chaubé* they had a treasure, and they made it difficult to obtain a specimen plant. Rauwolf attempted to spread the good news, but apparently the time was not yet ripe. It was not until the reign of Louis XIV that we find coffee, previously suspected of being a most dangerous drug, firmly established in the Western way of life. After this trip Rauwolf returned to Germany where he continued his botanical research.

Pierre Richer de Belleval was also a pupil of Rondelet. He was a respectable professor, but he intrigued everyone with the notion he never tired of repeating: "We should have a botanical garden of our own. The only means of achieving continuity in our work is to see the plants every day, all the time." The end of the sixteenth century was to witness the fulfillment of his dream. Thanks to the intervention of the duc de Montmorency, governor of Languedoc, Henry IV authorized the establishment in Montpellier of a garden designed to house collections of "living plants, permitting students to acquire visual knowledge of the herbs and plants necessary to them." The garden opened in 1593. The first director was Pierre Richer de Belleval. One part of the garden was reserved for herbs and medicinal studies; another, for attempts at acclimatization, in posthumous homage to Pierre Belon. The Montpellier Gardens, which today boast almost eight thousand species, contained then, as now, the juniper tree (*Juniperus*), the turpentine tree (*Pistacia terebinthus*), the lentisc (*Pistacia lentiscus*), and the cistus, all beloved by the first traveling botanists.

Botany, like many other sciences, reached a major turning

point in the sixteenth century. Numerous books appeared, and the pharmaceutical use of plants was greatly expanded, gaining in prestige. At the same time, the process of acclimatization was officially launched. France certainly had to make up for much lost time, particularly with regard to Italy, where the first botanical gardens had opened some time ago—Pisa in 1544, and Padua, established by a decision of the Venice Senate, in 1525. In Italy also a new job had been instituted, that of "gardener-acclimatizer," a position that was to be an important one for botany in general. When Henry IV entered Paris in 1594, after his coronation at Chartres, the town had as yet no botanical garden. Some garden planting admittedly took place here and there in the capital, as when the apothecary Nicolas Houel in 1577 created a herb garden at the Maison de la Charité Chrétienne. Most importantly, a horticulturalist, Jean Robin, possessed a "very distinguished" garden on the point of land near Notre-Dame. Robin grew herbs for the king's doctor and, in 1601, received the title of "arborist, herbalist and botanist to the king, curator of the garden of the faculty."

Robin was helped by his son Vespasien, who was born in 1579 and who loved flowers as much as his father. Vespasien traveled to Germany, Italy, and Spain and met there navigators and missionaries who "had been there." The young Robin made contacts everywhere, purchased seeds and bulbs, and soon the Robins' garden could meet the Parisians' demand for ornamental plants, the latest craze. Vespasien was particularly successful with bulbs, hyacinths, tulips, irises, and tuberoses (*Polianthes tuberosa*), but also with other plants: doubled-flowered anemones, ferns like the maidenhair (*Adiantum*), asarum, and the geranium. He watched with particular attention over the introduction of the *Robinia pseudoacacia*: the black locust. A specimen was brought to him from America. Most of the locusts growing today in Europe come from that single tree, named for him. He planted it in 1601 in what is now the

Square Viviani, by Saint-Julien-le-Pauvre, where it still blooms every spring.

Robin corresponded with, among others, a most chivalrous gentleman, Nicolas Claude Fabri de Peirsec, councillor at the parliament of Aix-en-Provence. He was perhaps the first person to set a collector's mind to work on ornamental gardens. This amateur had transformed the gardens of his château of Belgentier into a bold, exotic paradise. He had visited Italian gardens, particularly that of Padua, and learned there how to acclimatize, among other new plants, the Indian jasmine (*Jasminum humile*), yellow and highly prized; the Egyptian papyrus (*Cyperus papyrus*); the lifa of Mecca; rare vines of Smyrna, Sidon, and Damascus. He germinated a nutmeg (*Myristica fragrans*) and grew both ginger (*Zingiber officinale*) and the yellow-flowered American gelsemium (*Carolina jessamine*). To these can be added the medlar (*Mespilus*), the stoneless sour cherry (*Prunus cerasus*), and the banana (*Musa sapientum*). In Marseilles, Peirsec had contacts among ships' captains, who brought him back seeds, but none of his activity was official, or even organized.

Vespasien Robin's efforts were also followed with passionate attention by a brilliant man of universal curiosity, a famous doctor named Guy de la Brosse. He, like Richer de Belleval, whose establishment in Montpellier he had visited, believed that it was essential to provide the capital city with an immense garden in which might be brought together all the scattered treasures and where men could be trained to acclimatize new discoveries. But when Henry IV died in 1610, the project had still not been realized. It was Louis XIII who decided, in January 1626, at the suggestion of his chief physician, Jean Héroard, a friend of Guy de la Brosse (who had meantime become physician-in-ordinary to the king), to create a royal plant garden, a decision recorded by Parliament on June 6, 1626.

Jean Héroard was appointed superintendent of the future

royal garden, and before his death in 1628 he named as director his friend Guy de la Brosse. It took another seven years for the organization to get under way and for a suitable site to be found. Arrangements were completed in 1635. The site chosen was the Clos Coypeau, an estate that belonged to the heirs of the Daniel Voisin family. The Clos Coypeau was purchased for 67,000 livres, according to the deed signed by Maître Cornuel, the notary. Named the King's Garden (*Jardin du Roi*), the park finally opened its gates in 1640. Guy de la Brosse had supervised every detail of the planning and construction. He decided the planting areas, the future research laboratories, the teaching to be carried out, and everything that would make this place a meeting point for botanists and other researchers in natural science who might come from all over the world to consult their French colleagues.

Naturally, Guy de la Brosse called upon Vespasien Robin to assist him. Medicinal plants were to retain their rightful place in the King's Garden, but Robin also introduced luxury and exoticism. The brilliantly clever Robin himself set up the conditions necessary for successfully acclimatizing the most delicate ornamental plants and flowers. Long before his death in 1662, he witnessed the successful outcome of some of his endeavors.

From the late seventeenth century onward, the French botanical school was considered one of the best in the world. With both the Montpellier and Paris gardens available, traveling botanists who deposited their precious cargo could be sure that the best possible use would be made of it. While pure research continued for the most part to be performed at Montpellier, Paris fast became the chief center of attraction for all plant lovers. A capital city is always more eager for novelty and colorful display.

Guy Crescent Fagon, supervisor of the King's Garden. (Photo by J.-L. Charmet)

Father Schall in China. (Paris, Bibliothèque Nationale. Photo by ERL)

[4]

The Seventeenth-Century Missionaries

THE missionaries who had been the first to observe exotic plants in faraway lands were more interested in saving souls than saving flowers. During the sixteenth century, however, they had made their own contribution to plant collection. One of the most important was that of Father André Thevet. André Thevet, who was born in Angoulême, was a peaceful man to whom nothing would ever have happened had not Cardinal Charles de Lorraine, in 1588, conferred upon him the role of almoner on a ship loaded with Huguenots going into exile in Brazil. The task assigned to the most Catholic Father Thevet consisted of facilitating relationships between the equally Catholic crew-members of the vessel and its Protestant passengers. Apparently the priest was successful, for everyone arrived safely in port, on the island of Ganabra in the Bay of Rio.

Once on land, Father Thevet roamed far and wide. Soon he was attracted by a plant the local inhabitants called *petum*—tobacco. What caught his attention was the odd use made of this plant, which the natives first picked with great care, then dried in the shade. When dry, they wrapped it in a sticky reed leaf, placed the resulting object between their lips, lighted the other end, and breathed in. Smoke came out of the nose and mouth. It was a curious

31

custom, but those who followed it claimed that the plant was healthful and that when used in this manner it effectively dispersed and
consumed excess humors of the brain, even "dispelling hunger and
thirst for a time." When he observed that some of the westerners
accompanying him were extremely partial to the leaf, André Thevet
hastened to partake of it. "At first," he reports, "there is some danger
in this habit, until one becomes accustomed to it, for the smoke
causes sweating and weakness, even fainting fits, as I myself experienced."

This is the first French text about tobacco (*Nicotiana taba-
cum*). Father Thevet did not confine himself to writing about it.
He brought back seeds which he grew in his gardens in Angoulême.
Then the French ambassador to Lisbon, Jean Nicot de Villemain,
became involved. He had been offered tobacco in Portugal, where
the use of the plant was already beginning to spread, as it also was in
Spain. Better initiated than the father from Angoulême, Nicot was
the first to introduce the plant to the French court, and soon the
habit spread to the rest of the country. At the time it was called
"nicotiane." It was much prized when ground into snuff powder.
Others liked to chew it. Only in the nineteenth century did the
French begin to smoke tobacco like the Indians. Despite its high
price, tobacco quickly became so popular that in 1629 Richelieu
decreed it a royal prerogative, controlled its importation, and levied
a heavy tax on all sales.

In the seventeenth century botany became a real preoccupation
of the missionaries, mainly the Jesuits. Their quest for new plants
was sustained by the Court's passion for flowers. The Thirty Years
War raged for most of the first half of the century. It was a long, terrible, and disruptive struggle. The French began to look at nature
more fondly, with new eyes, as a welcome relief. The king's
brother, Gaston de France, duc d'Orléans, talked endlessly of his
garden at Blois. He invited artists there to paint the blooms "on the
spot." Nicolas Robert from Blois and Claude Aubriet did their best.

Suddenly, landscapes were no longer the farthest points in the background of pictures, but themselves bcame the theme of works of art. Mannerism and the cold mythological style of painting disappeared. They were replaced by a new realism later more fully developed by the Le Nain brothers and Georges de la Tour. For the time being, even tapestries were overflowing with flowers, as were gardens, vases, and canvases.

The first important discoveries of those years were made by Father Gabriel Sagard who, in 1632, was traveling through Huron territory in Canada. He was not the first to go there. Jacques Cartier and Samuel de Champlain had already passed through the area and French explorers had brought back many plants. But Father Sagard gave details of the vegetation of Quebec. There were roses, fruits, and roots there, as everywhere else, but there was also something the Hurons called "tortoise shell," the leaf of which, apart from its color, reminded Sagard of a lobster shell. One could drink from it "the dew which falls into it every summer morning." It was a carnivorous plant, the sarracenia (*S. purpurea*). The father also described how the Indians made sunflower oil, and discovered the sugar maple, from which was obtained the famous syrup. The procedure was quite simple: "We made a groove in the bark of this tree, then, holding a bowl beneath it, caught the juice or liquid which ran out. It helped us get back on our feet again when we were unwell."

Farther south a Dominican, Father du Tertre, explored for sixteen years the West Indies where he had come in 1635 with a naval expedition. Throughout this time he struggled to send to the King's Garden samples of his discoveries. But transportation was uncertain and Father du Tertre undoubtedly sent off many more specimens than those which reached Paris. Nonetheless, it was from him that Vespasien Robin received the maidenhair ferns (*Adiantum*), nowadays among Europe's most delicate house plants. The father declared that he had never seen a plant with "so much delicacy and refinement." It had moreover, a medicinal use. Father du Tertre saw the natives treat arrow wounds and snake bites with it, and women used it as a fertility aid. He also discovered a strange thistle, which he described as best he could and sent a specimen back to France. The plant was a cactus. Nor was this all. There were also the amazing island woods, which the Dominican was at a loss to describe scientifically. He sometimes spoke of rose, sometimes of red wood, then again of ironwood, never tiring of the subject. For this reason, his work played an important part in encouraging others to send full-scale expeditions to the West Indies to fetch unusual wood.

We come here to one of the most dramatic aspects of our story—these boats, so slow, so often sunk, combined with the problem of the unopened or mislaid crates in Paris. Indeed, one of Father du Tertre's crates, which waited too long in some shed, contained a small cutting of a shrub called the cacao tree (*Theobroma cacao*). The inhabitants of the West Indies liked to crunch the seeds of its fruit, which were also delicious when toasted. If the crate had been opened in time, the Spaniards would not have held for so long the monopoly of the chocolate trade.

Given such misadventures, it is understandable that the King's Garden was not yet as rich as it might be. One man, however, was to rectify the situation. In 1638, the same year as Louis XIV, Guy Crescent Fagon was born. He belonged to the first generation of Parisians for whom the King's Garden was filled with childhood memories, all the more so for him as he was the nephew of Guy de la Brosse, first director of the Garden. Every corner of the Garden soon became familiar to the young Fagon, who was fascinated by the vegetable world. His future course was clear. He was a brilliant student, and he spent every spare minute in the Garden observing the trial grounds and watching for new arrivals. At first he was doctor to the queen, then to the king (Louis XIV). He relinquished his Chair of Chemistry in 1664 in favor of one in Botany—at the

Garden, of course, where he was also appointed superintendent, a post he occupied until his death in 1718 at the age of eighty.

When he took charge, Fagon discovered that the Garden concentrated too much on small acclimatizations of no consequence. Daily activity there had become constricting and monotonous. The lost cacao tree was but one example among many of what must not be allowed to happen again. He was also disturbed by the Garden's policy of specializing in medicinal plants and emphasizing short-term returns. The initial enthusiasm had faded—even Vespasien Robin, that most skilled of gardeners, seemed overwhelmed by the sight of all the unfilled space. The same decline was occurring at Montpellier. The garden was destroyed during the 1622 siege of the town, but the plants were saved under the cover of night by the loyal Richer de Belleval. In 1629 Belleval began again, with a program of plantings to be spaced out over several years. But this had been implemented only slowly. The botanists felt as if they had been marking time for over thirty years.

What, then, had become of the glory France was to receive from her botanical gardens, intended to outshine those of Italy? Certainly the French were not unenterprising or lacking the taste for adventure. The problem had always lain with their methods. It would still be easy to hunt for new plants: Fagon was well aware that Louis XIV would help him send botanists all over the world. Even the formidable Colbert, keeper of the purse strings, would support such a venture. The difficulty was in acclimatizing the plants. A coffee-plant seedling provided the answer. One day, Fagon's assistant, Sebastien Vaillant, a talented botanist, received a coffee seedling from Leyden. In order that it might survive, he made, following the suggestions of his Dutch friends, a small glass house "tall enough for a shrub." Inspired by this, Fagon immediately ordered built not a small house but vast ventilated glass buildings, heated by pipes. Greenhouses were born. Tropical plants could now come to France; the small coffee plant grew tall.

Now men had to be found to set off overseas again. Here also Fagon was to be fortunate. Some missionaries were still determined to perform botanists' tasks. Notable among them was Father Plumier, who was born in 1646 in Marseilles. He entered the order of Minims at the age of sixteen and at first he studied mathematics and drawing, for which he soon proved himself remarkably gifted. In Rome, while finishing his studies, he attended the botany courses of Father Philippe Sergeant, and these proved to be an inspiration to him. In plants he found a better world here below, and he decided to devote his life to them. Plumier met and spoke at length with the great Sicilian botanist Boccone, to whom he soon dedicated a plant of the papaveraceae family, the Bocconia. Back in France, he gathered plants in the Alps and in Provence with the young Tournefort, who was making his first botanical expedition there. Plumier assembled a sizable herbarium, complete with a number of drawings. Time went by until one day in 1689 his chance came: He was invited to leave for the West Indies with Surian, a Marseilles doctor and apothecary. This, the first of what was to be four voyages, earned for Plumier the title of "the king's botanist," conferred upon him by Fagon.

The West Indies, where every plant seems enormous, even excessive, fascinated Plumier, who drew everything he saw, reconstituting entire botanical scenes in order not to isolate his specimens from their environment. He sought to portray the splendid spectacles provided by nature in these lands. Above all, he loved the West Indian ferns, which are several meters high. "Of all the plants I have discovered in the islands of America," he wrote, "scarcely any have given me as much pleasure as the ferns." He drew and described almost two hundred species of fern, 102 in Haiti, 63 in Martinique, and 32 in Jamaica. From the ferns Fagon established in the greenhouses of the King's Garden descend most of Europe's indoor and winter garden ferns.

Plumier drew endlessly, leaving behind him almost six thousand drawings, mostly line sketches, but many in color. Perfect as the ferns were, they were not the only plants of the islands. In the West Indies alone Plumier named, described, and drew 705 new species. We do not now have his herbarium which went down with the ship transporting it, but his texts—more than twenty volumes, including the splendid *Traité des Fougères*—and his drawings testify to the frenetic activity of this strange missionary.

Soon the rumor was to spread in Rome of a miraculous powder from Peru. It was quinine, popularly known as "Jesuits' powder" because it was brought back by a Jesuit, Father Tafur, at the turn of the century. It was produced from the cinchona tree (*Cinchona officinalis*), which no one in Europe had ever seen. Fagon and Louis XIV, intrigued, immediately decided to send Father Plumier in search of the plant. But his earlier travels had weakened him, and at Cadiz, while waiting for the boat for Peru, he caught a bad cold or pleurisy, and he died in 1704. The cinchona tree would have to wait until the time of La Condamine and Joseph de Jussieu to be brought back, but Tournefort did later give Plumier's name to an extremely fine variety of frangipani, the *Plumieria rubra* from tropical America.

In China the Jesuits gave the clearest proof of their abilities as missionaries as well as botanists. In the seventeenth century they were already well settled in the Celestial Empire. They knew the history of the country better than many mandarins, spoke the language, and, because they possessed a remarkable knowledge of Western culture, were able to amaze and delight the Chinese emperors. Those rulers did not hesitate to surround themselves with such adaptable and mild-mannered men, who loved, as did the emperors, to spend long hours meditating in beautiful gardens. An arrangement was reached with regard to religion. The missionaries allowed the Chinese Christians to call God *T'ien* (Heaven) and to retain ancestor worship. The compromise was working excellently. Several thousand Chinese had converted to Christianity, the Catholic religion had been officially recognized, and churches were being built. Father Schall, a German Jesuit, had become counselor to Emperor Chouen-tche, then tutor to his son, the young K'ang-hi.

Another Jesuit, Father Etienne de Fèvre, alerted French botanical circles to what had been going on in China since time immemorial. The French learned that the art of flowers and trees was sacred in China, that it had attained an astonishing level of refinement, that medical plants had been appropriately classified since remotest antiquity, that the emperor's gardens were maintained by mandarins who watched over the bamboo with particular care, that the tree of wisdom was the fig tree, the elegant Gingko was the guardian of pagodas, and that an astonishing plant, the ginseng (*Panax schinseng*), conferred eternal youth on those who knew how to use it. (The Chinese ginseng is a cultivar of *Panax quinquefolium*, the American species.)

This report created an uproar in Versailles. Little was known about China, except that it had between 100 and 115 million inhabitants and produced porcelain and silk, which were traded through the ports of Canton and Macao. The news was exciting. The chance soon came to learn more. Father Schall had just died, but his successor, Father Verbiest, was equally well received in China; the new emperor, K'ang-hi, had named him director of the Peking observatory. Verbiest, a friend of Colbert, wrote to the king in 1685 asking for new assistants. Louis XIV acquiesced immediately, designating ten Jesuits, who were to be sent to Peking at once. Their departure was celebrated by a grand reception at the Académie des Sciences. Six of the men were to reach China by way of Siam and the sea while the four others were to travel overland. This expedition, known to historians as the "Peking Mission," took two years to reach China and arrived in 1687. Its chief members were the fathers Guy Tachard, Louis Le Comte, d'Entrecolles, Parennin, and d'Incarville. They set to work immediately, soon providing the Académie des Sciences with reports and Fagon with seeds.

Father Dominique Parennin accompanied the emperor on trips to Manchuria. He was the first to send wistaria (*Wisteria chinensis*) to Europe. Then came the Chinese pink (*Dianthus chinensis*), followed by a flower sent by d'Incarville, the China aster (*Callistephus chinensis*), a sumptuous blossom with dense petals, the seeds of which produced varied and unexpected specimens with double or triple flowers. The King's Garden also received the first Chinese delphiniums (the perennial *Delphinium grandiflorum*), the first Chinese rhododendrons, and "Asian jasmine" (*Jasminum officinale*), with flowers so delicately scented it could be used to flavor tea. Tea (*Thea sinensis*) itself finally reached France, where attempts were made to grow it under glass with only mediocre results. Because there was no demand for tea at that time, the project was abandoned. The French began to drink tea only when Anglomania struck France in the late eighteeth century. For the time being they were more interested in the soy bean (*Glycine hispida*), the lilium, the ailanthus (*Ailanthus altissima*), the imperial peonies, the hibiscus, the paper mulberry (*Broussonetia papyrifera*), and the oriental thuja, with a more scented wood than any other known in those days, so much so that it was burnt in sticks, like incense. More fabulous plants were coming: the spindle tree (*Euonymus*), the forsythia, whose yellow flowers appear before the leaves, the hydrangea, the gardenia, and the camellia.

The French could at last study the ginseng. Everyone was talking about that plant, both at court and in the King's Garden. Since it was seen in China as a true elixir of long life and restorer of virility, the French called it "panacea." The root of this plant of the Araliaceae family contains elements so rich and varied and dangerous—the Jesuits recommended the utmost caution in its use— that only the Ecole de Pharmacie was authorized to analyze its peculiar properties and medicinal possibilities. Today ginseng, which has never been successfully acclimatized in the West, appears in several medications because it contains the natural equivalent of certain hormones.

The French also learned that the Chinese were past masters in the art of dyeing fabrics, obtaining blue from indigo (*Indigofera tinctoria*) and red from madder (*Rubia tinctorum*) and safflower (*Carthamus tinctorius*). Yellow, the color most frequently used in China, was derived from the orange fruits of the *Gardenia jasminoides*, from *Garcinia cambogia*, from a kind of mignonette, and from the flowers of the saffron crocus (*Crocus sativus*). Only the turmeric rhizome (*Curcuma longa*) yields kiang-hoang, the imperial yellow. But, to Fagon's distress, once in the King's Garden, these plants mysteriously lost their coloring power.

Bamboo (*Bambusa*) was also intriguing. Years had already been spent trying to understand its almost metaphysical life, its enigmatic germination, the mysteries of the mother plant, and the overflowing life hidden beneath its smooth, lovely exterior. Bamboo was used in China as we use plastic today. Everything was made of bamboo—paper, chopsticks for eating rice, boat sails, ropes, and even waterpipes. The Jesuits told Versailles about fabulous, ancient Chinese gardens in which everything is reduced, even great trees, lakes, rocks, and rivers. Gardeners there indulged in delightful caprices, even changing the color of flowers and the shape of petals, and reportedly crossing the chrysanthemum and the artemisia, grafting everything into everything, oak onto chestnut, peach onto persimmon (*Diospyros kaki*), and obtaining fruit from a quince (*Cydonia vulgaris*) crossed with an orange (*Citrus sinensis*). The Chinese apparently did anything. Could not the Sun King do as much?

No. Louis XIV was preparing to send for some Chinese gardeners, but in 1704 history recalled itself. There was great concern in Rome at the concessions the Jesuits had made to Chinese philosophy. The Pope, Clement XI, issued a decree condemning the

Chinese rites. If China would not be Roman Catholic, it must be forgotten. The Jesuits were reprimanded. In vain they warned that a strict Roman Catholic rule could not be accepted in China. No one listened. And so the Emperor K'ang-hi closed the gates of China to all Europeans not bearing diplomatic letters. At the same time, however, he asked the missionaries to draw the Great Map of the Chinese Empire. K'ang-hi, who was beguiled by the missionaries, died in 1722. His successor, Yong-Tcheng, prohibited Christianity in 1724. The Chinese dream ended. Versailles now had rhododendrons, jasmine, and sweet oranges, but these provided scant consolation. The curtain had fallen over a world of botanical treasures.

A meeting of scholars in the Garden drugs room.
(Paris, Bibliothèque Nationale. Photo by ERL)

Louis XIV, Colbert, and the Court visit the Garden.
(Paris, Bibliothèque Nationale. Photo by ERL)

[5]

Tournefort *in the Levant*

Joseph Pitton de Tournefort, later known simply as Tournefort, was born in 1656, in the family château of Tournefort, nearly nine miles from Aix-en-Provence. His father, Pierre Pitton de Tournefort, had married a young woman from Aix. They had nine children, seven girls and two boys. Luc, Joseph's elder brother, inherited his father's position as *secretaire du roy*. Joseph had a warm and happy childhood in one of the most beautiful parts of France. When he was very young, he was influenced by the ambiance of Aix, which at that time attracted many well-educated men, a number of whom held botany as a favorite occupation. They thought of it as a kind of recreation. One of the most remarkable among these Aix truants had been Nicolas Claude Fabri de Peiresc, that magistrate enamored of ornamental gardens, who had his own at Belgentier, the equal of the finest in Europe, and, as we have seen, had been the friend of Vespasien Robin. In Peiresc's circle, plant study had almost become a philosophical experience. People came to botanize and to listen to the experts, who were not necessarily professors but often just local inhabitants. The university of Aix, while famous, was still overshadowed by Montpellier. The chief area of specialization at Aix was law. A chair of botany, in these middle years of the seventeenth century, was extremely difficult to establish. Those eager to study natural science were still well advised to go to Montpellier. But, when they were there, something subtle and indefinable marked them and bound them together— they were "*Aixois.*"

Determined to provide a solid education for his son, Tournefort's father put him under the care of the Jesuits, with whom the young Joseph learned ancient languages, physics, chemistry, anatomy, medicine, and philosophy. As a younger son, he was supposed to enter the seminary, and he would have done so out of a sense of duty had his father not died in 1677. Tournefort, now twenty-one, could then abandon himself to his passion for natural science. Soon he left Aix for Montpellier, where he was to become a botanist under the guidance of the master Pierre Magnol, who taught there. Magnol later became Louis XIV's doctor, and the magnolia was named for him.

In Montpellier botany was considered an advanced science and it was well established. As soon as he arrived, Tournefort plunged into the study of the chemical life of plants and the analysis of their organs, a common practice there. Later he became interested in agronomy. A brilliant, scholarly man, the young Tournefort was also exuberant and elegant, a man in love with life. A refined epicurean, he surrounded himself with agreeable friends who, like him, preferred knowledge over pedantry. Education took place amid the comforts of life—fine wines, pretty women, and opulent gardens. But let us not be deceived by these superficial signs of a *grand siècle* education: Beneath them, Tournefort had a determined character and was as rigorous in his research as he would be strong in adventure. He worked hard, inventing a new plant classification, an area still full of confusion. "We must therefore," declared Tournefort, "apply a precise method to the naming of plants, for fear that the number of names does not eventually equal the number of plants. This is exactly what would happen if everyone could name each plant as he wished. The result would be not only great confusion but an astonishing burden for the memory, weighed down by an infinite number of denominations." The point was well taken, but where could he begin? What Tournefort most loved about a plant was its flower. Why not begin there, classifying plants according to their flowers, after careful examination of the different parts: petals, calyx, stamens, and pistil? "There is often," writes Tournefort, "one great difficulty in this matter, namely, the distinction between flowers with and flowers without a calyx. Among botanists, it is agreed that roses, carnations, pear, and cherry tree flowers have a calyx, but nature has not bestowed one on the lily, the tulip, the hyacinth, and many other plants. One wonders as well whether the foliaceous part of certain flowers should be called calyx or corolla—for example, with the orache (*Atriplex hortensis*), the lady's mantle (*Alchemilla*), and the asarum." Tournefort began to study closely the nature of flowers and he used his findings as the basis of the main axes of his classification system. He established what was subsequently to be called "a complex structure of constant characteristics." He founded the first botanical *system*. This represented a considerable advance for the period and, of course, Fagon's interest in Tournefort was aroused.

Fagon summoned Tournefort to Paris and named him professor of botany at the King's Garden in 1683. During this year Tournefort worked on several books and conducted experiments aimed at explaining the mysterious fertilization process of the palm trees he had brought back from a trip to southern Spain. He had also just completed his masterly *Eléments de botanique ou Méthode pour connaître les plantes*, a work composed in Latin but immediately translated into all European languages and into popular French. The book inventoried and studied 8,846 plants known at the time. It was a huge success, and in 1699 Tournefort was elected to the Académie des Sciences. A profitable official career and a peaceful life as a great scholar lay before him when, suddenly, Louis XIV and Fagon proposed that he make a long journey to the Levant.

Everyone wanted Tournefort to journey from Marseilles to Mount Ararat. Fagon because, like Tournefort, he was haunted by the mythical names the ancients had given to plants and wished to make a science out of these myths; the secretary of state in charge of the navy and the academies, the comte de Pontchartrain, because he needed verification of the first coastal maps then being sent to his ministry; and Louis XIV because he wanted information concerning the Levant, about which his ambassadors had but meager knowledge, for each trip they made into the countryside was hemmed in by countless formalities. Tournefort accepted the proposal with enthusiasm. He described his mission in the following terms: "My lord the comte de Pontchartrain, secretary of state charged with responsibility for the academies, and always attentive to what might benefit the sciences, proposes to His Majesty at the end of the year 1699 to send to foreign countries people capable of carrying out observations there, not only of natural history and ancient and modern geography, but also of the trade, religion, and customs of the various inhabitants." The official mission brief added that Tournefort was to undertake "the identification of the plants of the ancients, and perhaps, in addition, of those they missed."

Tournefort's compatriots were to be full of enthusiasm for this journey. In many respects it was a pioneering one. The traveler captured contemporary imagination by the mere fact of his itinerary. Well before the scientific results were known, his *Relation d'un voyage au Levant* appeared, which described "the Islands of the Archipelago of Constantinople, the Coastal Regions of the Black Sea, Armenia, Georgia, the frontiers of Persia and Asia Minor, complete with maps of the towns and important places." It was a resounding success.

On April 13, 1700, Tournefort left Marseilles with his two main collaborators: Claude Aubriet, the draftsman, and André Gundelscheimer, a young German botany student who was also trained as a doctor. His training would prove useful to the small expedition. Tournefort's first decision was to travel overland as much as possible in order to have plenty of time for botanical research. But he also took another precaution: To facilitate matters on his return, he asked his correspondents, Fagon, Pontchartrain, and the Abbé Bignon, secretary of the Académie des Sciences, to preserve carefully all the letters he wrote to them. He soon realized, however, that this was not so simple as it appeared. Those to whom he entrusted his letters were often assassinated en route. He was compelled to revise his system. He continued to write to his friends but he also took notes which he kept in his own possession.

The first stop was the Greek Islands. When they climbed Mount Ida, Tournefort was less enthusiastic than Pierre Belon: "With all due respect to Jupiter," he declared, "who was, they say, raised and buried here, this is the most unpleasant mountain I have ever seen. It looks like the bare back of an ugly donkey, with no trace of a forest, no landscape, pleasant solitude, stream, nor spring." On Delos he was struck by the pillaging of the temple: "The masons from the nearby islands all come here as if to a quarry, to choose the pieces they need. A beautiful column is broken to make a stairstep. Turks, Greeks, and Latins break, overturn, and carry off whatever takes their fancy." The three friends wanted to go to Athens, but they had to abandon the idea. "It is very dangerous to make the trip to Athens at the moment," Tournefort explained. "The countryside is full of ill-disciplined troops who attack and rob travelers, often dispatching them to the next world. As we are well pleased with the present one, we avoid these kinds of people."

After Greece they reached Constantinople, an amazing city: "It seems as if the Dardanelles and Black Sea canals were created in order to bring to Constantinople the treasures of the entire world." Later he wrote two volumes devoted exclusively to the customs, religion, and society of the Turks. From Constantinople they set off for Asia, feeling for the first time a real sense of danger and adventure. The roads and people were now less familiar. For the journey to Erzurum, however, Tournefort was fortunate in being able to join the caravan of the pasha of that town, Numan Kupruli. Tournefort described this stage of the trip: "We left Constantinople on April 13, 1701, in the train of one of the finest gentlemen in the world, perhaps the only person in Turkey deserving of the epithet— the pasha of Erzurum. He is a tall man, cold as ice but of sound judgment, devoted to study, particularly of Arab works, a skilled politician, well aware of the interests of the princes of Europe, but an avowed enemy of every kind of injustice, robbery, and affront."

The pasha had an impressive caravan—three hundred camels, six hundred people, and many surprising sights. "All this afforded us a fine spectacle. The women were carried on camels in litters curved like a crib. The tops were covered in printed cloth. The rest was barred on every side with more care than the parlors of the most austere nuns. Some of these conveyances looked like cages perched on the back of a horse or camel." Contemporary readers loved these anecdotes.

Stopping to gather plants on the edge of the Black Sea, they were watched with amazement by the other travelers. "Every day we would run like urchins along the shores of the Black Sea, the most beautiful coastline in the world, so covered in woods that from afar it appears black, the reason, perhaps, for the name of the sea."

They halted again at Trabzon, where they spent a month. Tournefort gathered quantities of unknown and unexpected flowers: Nature had here completely regained control. Tournefort noted that "Trabzon, once the capital of a mighty empire, is now a most mediocre city. The marble palaces and porticoes have been destroyed, but nature has remained completely intact." They set off again, the pasha's musicians playing in time with the slow pace of the camels. On June 15 they reached Erzurum, where Tournefort and his companions bade farewell to the pasha and his beautiful prisoners. They gathered plants in Kars and proceeded to Yerevan, where the patriarch of Armenia received them with great professions of friendship. Next came the ascent of Mount Ararat, the final stage of their journey. Climbing up excitedly, they were convinced the mountain must be covered with marvels, since they had traveled so far to see it. Disappointed, they realized that many stories had been invented about this inaccessible spot, "where, it is believed, perhaps wrongly, that Noah's Ark came to rest after the Flood. Everything published by travelers about the mountain is false. There are neither monks nor hermits, neither solitaries nor recluses. It is a horrible mountain, as high as our Alps. Half of it is hidden by snow all year round. The rest is bare and sandy, covered with tragacanth (*Astragalus tragacantha*) and small juniper trees. One sees many partridges and a few tigers." Scant supplies here for the King's Garden. They decided to begin the journey home.

Tournefort, however, wished first to visit Georgia, which delighted him, and to botanize around Tiflis. He had three reasons to go there: "first, to see this land, of which so many marvels are told; second, to collect seeds from the many beautiful plants we had only seen in bloom in Armenia and would have lost if we had tried any other method; and third, no less important a reason than the others, because we wished to wait until the roads of Anatolia, which we had to travel to reach Smyrna, were freed of robbers." It becomes clear that, from every point of view, Tournefort conducted his affairs with infinitely more prudence than Pierre Belon, the first botanist traveler. Times had changed.

After returning to Smyrna with a caravan, as usual, the group made a detour to Olympus. At last came an ascent not followed by depression: "We reached the top of Olympus, insofar as the snow permitted," wrote Tournefort to Fagon. "This part is covered with firs (*Abies*), and we found there some lovely plants, among which I particularly admired the true black hellebore of the Ancients, quite different from the hellebore given that name in Europe. You will be receiving a seed." The black hellebore was probably *Helleborus olympicus*, as opposed to our *Helleborus niger*, which is white.

Tournefort did, indeed, succeed in sending seeds to Paris by making arrangements with caravaneers and embassies.

Must he go to Egypt? He should, but there was an outbreak of the plague there, and Father Aubriet was suffering from troublesome headaches. The group also had financial problems: Travel and official missions were all very well, but the administration had to show support. "From the letter of credit arranged for me by M. de Pontchartrain," wrote Tournefort, "there remain only 3,680 livres, a sum I believe insufficient for completing both the trip to Egypt and the return to France, for in Egypt we shall have to make several caravan expeditions into Arabia and Ethiopia." The 1,500 livres he expected to collect in Smyrna never materialized. He begged his correspondent to refer the matter to the king: "His Majesty was good enough to set our expenses at 15 livres a day. At that rate I should receive a further 3,000 livres, but I think that 2,000 will suffice to complete our journey."

Another problem was the fact that all of them were, quite simply, tired. Tournefort and his companions were sated with images, and sad without news of Europe. One piece of information did reach them, however: Fagon was unwell. If anything should happen to him, with whom could they classify and analyze the material they had collected? They decided to return home. Dressed in Turkish style, Tournefort landed at Marseilles on June 7, 1702, after two years of traveling over hill and vale. Fagon, Pontchartrain, and the Abbé Bignon awaited him in Paris. His return was more glorious than that of Belon. Louis XIV received him and listened to the account of his journey. Tournefort had brought back over a thousand new plants, Aubriet had drawn some masterpieces, and the whole botanical world of antiquity had been illuminated. Wishing to honor Joseph Pitton de Tournefort, the king offered him the flattering position of doctor of his grandson, young Philippe V, now king of Spain. Tournefort refused because he preferred to continue his work in Paris. He spent most of 1703 in the King's

Garden, watching his seeds germinate and his young shoots grow.

There are two main divisions of Tournefort's work, one truly botanical and the other which relates to the new light he shed on the "drug" trade between East and West, a trade whose rules had before eluded analysis by Europeans. This latter contribution was of major importance, for it influenced the pharmaceutical "industry" and pharmaceutical "research" in all the Western countries. By *drugs* we mean the medicinal plants used for manufacturing powders, ointments, and the like. These drugs traveled first from the Far East to Smyrna and Constantinople by caravan. There they were delivered to merchants or adventurers who resold them in Marseilles or Cadiz, the two centers of the European trade. From these two places drugs moved on to the innumerable "druggists," or apothecaries, of the big cities.

Tournefort knew these caravaneers, who formed a marginal, disquieting world and an amazing itinerant universe. He soon noticed that precious Persian silks, for example, were but harmless low-profit merchandise in comparison to the mysterious drugs. "I can imagine nothing more difficult," he noted, "than to write a good history of drugs, that is to say, to describe not only their medicinal content, but also provide a description of the plants, animals, and vegetables from which they are derived. One would have to go not only to Persia but also to the mogul in India, which is the world's richest empire, where strangers are perfectly well received. Everything can be bought there."

In Tournefort's day, the land of the Great Mogul was in a state of upheaval, but it is true that centuries-old traditions of civilization were still respected. Caravans coming from the north of India would bring a mixture of brocades, cotton goods, and precious stones; poppies, spices, balms, and "all kinds of odoriferous resins" used in the manufacture of perfumes, pomades, and poultices. The caravans often transported laudanum, prepared from an extract of the poppy opium, mixed with saffron, cinnamon, or cloves.

But laudanum was already known in Europe. In addition, Tournefort thought one would find there aloes, different species of pepper
and myrrh, the resinous juice of which was used to treat pains in the
joints, as well as dragon's blood, the red resin of a palmlike tree
of the Dracaena family which made a homeostatic powder. Also
brought by camel were tea from China, coffee plants, and gum
arabic (*Acacia senegal*). The clandestine trading of dyes also interested Tournefort.

He was so impressed by what he had seen that, once back in
Paris, he hastened to complete an abridged history of drugs. He
revived Pierre Belon's dream of either acclimatizing the plants and
shrubs of the Terebinthaceae family, so as to be no longer economi-
cally dependent on the Orient, or else finding their equivalents
among French flora. By analyzing drugs, he wished to reduce the
mystery which enshrouded them. Sorcery was not dead in Europe,
but it was dormant when the Enlightenment dawned. Tournefort
well knew that many plants were as dangerous as they are beautiful.
Pierre Belon had said the same thing years ago, but again his words
were being repeated by this respected and widely recognized
scholar. The botanical gardens of Paris and Montpellier were immediately placed at Tournefort's disposal and their laboratories
mobilized to perform necessary research.

Tournefort's botanical contributions were immense. He was
most methodical in sending to Fagon specimens, berries, and seeds.
Because he was always fearful of shipwrecks and pirates, he kept
duplicates of all his finds, as is shown in a letter he wrote to Fagon
from Greece: "We have collected the seeds of 121 plants which I
have the honor to send you. I am taking this risk, even though many
vessels are being lost at the moment. I have held on to as many
again in order to still have specimens if anything should go wrong."

His prudence enabled Tournefort to bring back 1,356 plants
from the Levant, many more than had been catalogued by Dioscorides and Theophrastus. They included wild madder (*Rubia*

tinctorum), campanula, clematis, betony (*Stachys officinalis*), centaurea, the flower there called "Centaur's grass," plus marigolds (*Calendula*), honeysuckle (*Lonicera*), herb bennet (*Geum coccineum*), heliotrope (probably *Petasites fragrans*), rocket (*Hesperis*, either *tristis* or, more probably, *violacea*), cyclamen, lavender (*Lavandula*), ornamental thyme (*Thymus*), dragonhead (*Dracocephalum moldavicum*), mallows (*Malva*, which could be *M. mauritiana*), ranunculus (probably *asiaticus*), violets unknown in Europe, myosotis (*dissitiflora*), valerian (*Valeriana*), veronica, dittany (*Dictamnus*), and rock rose (*Cistus*) of Crete, arum, and artemisia (*vulgaris*), the *Cotyledon cretica*, or Venus' navel, dwarf cherries, and digitalis. He also brought back the gentian (*Gentiana pannonica* or *dinarica*), the woad (*Genista tinctoria*), hawkweed (*Hieracium zisphus*), new exotic irises, moonwort (*Botrychium lunaria*), and jujube tree (*Zizyphus jujuba*). Then there was the ephedra (*E. distachya*), a handsome woody plant from China, where it appears in the most ancient herbaria. The ephedra produces an active alkaloid, ephedrine, which is used in various medications for allergies. Tournefort was also interested in an ornamental shrub bearing clusters of shiny red berries. Eventually he classified it with the sorb trees (*Sorbus* family). In this way, the nettle tree (*Celtis australis*) entered the Garden.

We must mention next the acanthus (*A. mollis*), the plant fabled in Greek art, where it is preserved across the centuries in its unfading, sculptured beauty. Tournefort was the first to introduce it into France, as well as the orchid. His orchid was not, of course, the luxurious, indescribable flower of the South American jungles, but the one the Greeks called *orchis*, which grows as far as the west of Russia. The first specimens immediately took their place in the glass houses of the King's Garden.

Tournefort even sent back plants he knew to be poisonous. He wrote Fagon that it was important to draw the list of plants which were dangerous, and to do this promptly. As an example, he mentions the aconite (*Aconitum*, probably the deadly poisonous *A. napellus*), which he knew the ancient Gauls used to poison their arrows. In the Levant, the aconite was used both as a poison and an antidote, depending on the dose administered. He told the doctors of Paris about a most beautiful and dangerous kind of digitalis, the deadly nightshade (*Atropa belladonna*), which the Greeks called "the Poisoner." Belladonna is still used medically to dilate the pupils. Tournefort, with Fagon, finally set about classifying his marvels. But this last adventure, which should have been his true reward, Tournefort was not destined to complete. In 1706 he wrote his *Voyage au Levant*, not published until 1717, and was appointed professor at the College de France. On April 16, 1708, in the rue Lacépède in Paris, he was caught by the axle of a passing carriage and crushed against a wall. Like Belon, he died a violent death. Were all those who sought to penetrate the mysteries of the vegetable kingdom doomed to this fate? In his will Joseph Pitton de Tournefort bequeathed to the king his collections and two maples from Crete. They are said to be still living today. He asked to be buried at Saint-Etienne-du-Mont. The request was granted.

Death of Tournefort. "The doctor of botany and the doctor of mineralogy killed by a fashionable physician." Eighteenth-century engraving by "a connoisseur of botany." (Photo by J.-L. Charmet)

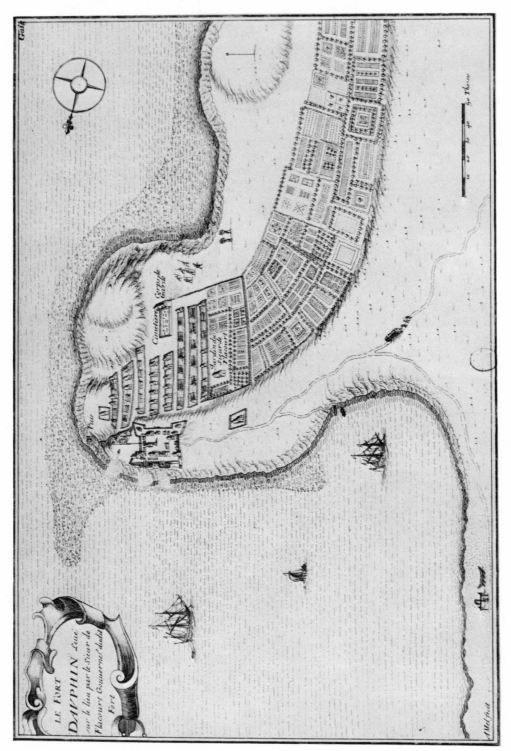

Trial grounds of Governor de Flacourt in Fort Dauphin, Madagascar.
Seventeenth-century engraving. (Paris, Bibliothèque Nationale. Photo by ERL)

Allegory of Mauritius Island. (Paris, Bibliothèque Nationale. Photo by ERL)

*The Botanical Garden of Nantes in the nineteenth century.
(Nantes, Musée Dobrée)*

[6]

Adventure in Canada

Versailles, the symbol of the Sun King's glory, was becoming the most magnificent estate ever conceived by a European monarch. Le Nôtre had surrounded the palace with admirable geometrical gardens that emphasized a new ideal of beauty. Rigorous design was seen as the only appropriate manner of expressing the aims of a great reign, and it was to become the hallmark of classical beauty. New trees and new flowers, brought back by the botanist-travelers and first acclimatized in the King's Garden, were planted there by an army of gardeners. Mansart had built the orangery to display fruit trees, oranges and lemons, from the Orient. Louis XIV was now turning his attention to the sea and to faraway lands, under the influence of his indefatigable minister Colbert. Colbert had set his mind on destroying the trade monopoly imposed on Western Europe by the Dutch and, to a lesser extent, the English. To achieve his purpose, he needed a powerful Royal Navy, merchant ships, and overseas trading posts. The adventures of one man, Etienne de Flacourt, in Madagascar were to precipitate events and cause France to become much more interested in the outside world.

The great island of Madagascar had not interested the English, Dutch, or Portuguese. The Compagnie française des Indes, however, had settled men there, creating Fort Dauphin in 1643 and calling the island "Dauphine." In 1648 the powerful company named Etienne de Flacourt commander of the island. Once there, Flacourt soon had problems, first with the natives because the island was divided into a number of small kingdoms, but also with his own men, who rebelled against him in circumstances that are still insufficiently explained. The Compagnie des Indes retreated before this chaos and left Flacourt stranded for six long years, bereft of boats, news, and everything else. At last he managed to send a message to France by way of a Dutch vessel. The company agreed to send him two boats. Flacourt's one wish was to return to France where he landed at Nantes on June 28, 1652. In 1653 he published *La Grande Isle de Madagascar*, a book in which we are told that the vegetation there makes the island a true paradise on earth.

What Flacourt wrote and, above all, what he brought back with him were enough to stir the imagination of the French botanists, who all declared that people must be sent there. Flacourt had returned with orchids known on Madagascar by the name of *tsingolo*. He had noted several dozen different species of these. He had also brought back gardenias, everlasting flowers (*Helichrysum*), tobacco, the jasmine Lalonda (said to be the "true" jasmine), raffia palms (*Raphia ruffia*), a new periwinkle (*Vinca rosea*), hibiscuses, and mimosas. He also mentioned, and here Colbert, eager to feed his people, must have pricked up his ears, the existence of a large number of edible roots. He deposited his discoveries at the botanical gardens of Nantes, which were by then beginning to play a significant role in introducing into France plants brought back by navigators. The *Jardin de Nantes* was created at the beginning of the seventeenth century. At first it was charged with producing "simples" to be used by ships' crews leaving the port. Later an order of Louis XIV obliged ships' captains to bring back to Nantes for "refreshment and comfort" plants eventually destined for transfer to the King's Garden. A privileged crossroads with a privileged climate, Nantes was enriched over the centuries with a rare collection of exotic plants and trees, some of which could not be acclimatized anywhere else.

Madagascar now became a universal topic of conversation. The French had to go there, and, while there, take possession of it for the Crown. The entire traditional scientific hierarchy was mobilized. The academies and everyone else became involved. Sailors prepared the first hydrographic reports of the area, the first coastal maps of the island were drafted, and the name *Madagascar* appeared on the sailors' atlases. The road was paved for future botanical explorers.

In France, meanwhile, three men turned their attention toward Canada: Louis XIV, who wanted to know how to assess the riches of this "province"; Colbert, who was in need of wood both for his ships and for replanting the French forests, now dangerously impoverished; and Fagon, of course, who had almost nothing from this vast American territory except thujas and yews from Jacques Cartier and maples from Father Sagard. Canada must contain more than that.

But since 1653 few trips had been made to the lands of snow. The West Indies were more in vogue, more exciting than that white country of vast and silent horizons, peopled by Indians the Europeans liked to think of as formidable. The Sun King set about arousing interest in the colder clime. He named as governor the comte de Frontenac, a man of spirit, whom he expected to reawaken the French taste for adventure. A fine soldier, Louis de Frontenac had pleased the king by defending Candia against the Turks. Since then, he had been living in retirement in his Touraine château. But there, this friend of Madame de Sevigné and of the principal artists and writers of the day, still dreamed of adventure. When Louis XIV had Colbert order him in 1672 to leave for Canada to make of it a large French province in which "justice and peace with the Indians" should reign, he leaped at the opportunity.

With him, New France, "*la Nouvelle France,*" as it was called, was to experience a period of real prosperity. The comte de Frontenac easily achieved an alliance with surrounding Indian tribes, mainly the Hurons. He succeeded in arousing such an enthusiasm that soon traders arrived with their families, together with scientists, geographers, and mineralogists. Only Fagon was heartbroken—there was not a single botanist in the bunch. Some men, however, were to come to his aid.

In 1670 Louis XIV and Colbert had decided to send a young adventurer to Nouvelle France. This young man, Robin Cavelier de La Salle, had every possible quality required in a good settler. First, he had an interesting past, despite his youth. He was born in Rouen in 1643 of a family of wealthy merchants and he was the nephew of Henri Cartier, one of the founders of the Commercial Society of New France. When La Salle left for Canada, he was only twenty-seven, but he had completed his studies; he was a brilliant mathematician, knowledgeable also in natural science. He had been a Jesuit, but had left the order. His restless mind needed new experiences. In Canada he bought a small farming estate near Montreal. La Salle's activities were closely watched from Versailles: It was hoped he would cultivate the Canadian soil and inspire others to follow. A number of emigrants, in fact, did join him, even before the arrival of Frontenac, and they founded families. This group provided the country with a whole new area of economic activity; agriculture was widely developed and trade became prosperous. Cavelier de La Salle did not forget Fagon. He sent him, to his delight, fine walnut trees, which served to renew the old French stock, ten mulberry trees, and eight new kinds of oak. He also sent the sarracenia, the strange carnivorous flower in which Father Sagard used to drink the morning dew. New thujas reached the Garden, and even a variety of ginseng (*Panax quinquefolium*), unfortunately not the good variety, which grows only in China and Tartary. These were followed by hawthorns, by a very fine vine, and an amazing cornflower (*Centaurea americana*).

But Cavelier de La Salle loved movement and activity, and was going to seek them for the greater glory of the king of France. Waterways then played such an important part in opening up the American continent that the comte de Frontenac encouraged the pioneers to find out where all the waterways led. For this reason, a Jesuit, Father Marquette, and Louis Joliet, both brought up on hunters' stories, undertook a successful journey to see the Great Lakes. When they returned, Cavelier de La Salle decided to go even further, and to go south. He organized an expedition to explore the Mississippi and traveled all the way to the Gulf of Mexico. En route he built fortresses here and there and formed the territory christened Louisiana, in honor of the king. So that all might hear of this exploit, La Salle sent to France some new orchids and a syringa (*Philadelphus*), which Fagon, on the king's orders, immediately planted at Versailles. If La Salle was to be believed, Louisiana was yet another paradise of sun and luxuriant vegetation. But it was as well a paradise full of snares—marshes and countless uncharted streams. It proved fatal to La Salle. In 1687, in mysterious circumstances, he was assassinated there by his companions. The task he began, however, was completed by another, Pierre Le Moyne d'Iberville. Accompanied by missionaries, sailors, and the first French immigrants, and assisted by pirates, d'Iberville set up in 1699 the first French trading posts in Louisiana. He also sent plants to the King's Garden, at the request of Pontchartrain, who alerted the scientific world to the need for mounting a full-scale exploration of Louisiana.

This was a splendid period for the Garden. In crates and entire boatloads, Fagon received the latania, or Bourbon palm, the persimmon (*Diospyros virginiana*), some cedars, heliotropes, and hyacinths; a lily of the valley (*Convallaria*), lychnis, sweet peas (*Lathyrus odoratus*), new ranunculus species, mignonette (*Reseda*) from the marshes, new dog roses (*Rosa canina* and *Rosa setigera*), splendid rhododendrons, a veronica (*Veronica virginica*), more and more syringas, almonds (*Prunus amygdalus*), sorbs, probably service berry or shadbush (*Amelanchier canadensis*), bananas, orchids, wild anemones (*Hepatica acutiloba* and *Anemonella thalictroides*), phlox (*maculata, paniculata, amoena, divaricata, subulata, ovata,* and *pilosa*), climbing bignonias (*Bignonia argyreo violascens*), giant geraniums—and cotton.

Even after Cavelier de La Salle had left Canada, the King's Garden was then still receiving plants from the colony. An army surgeon, Michel Sarrazin, who had come there to care for the soldiers, happened to be curious about botany, and he soon became familiar with the immense vegetable riches of Canada. Aware, however, of his poor botanical qualifications, Sarrazin returned to France to complete his education. He met and got on well with Fagon, and perhaps also Tournefort. He set off for Canada again in 1697 and thereafter sent regular consignments to Fagon.

His first package contained items of major importance, five species of birch: the "Providence Tree" (*Betula papyrifera*) and others (*B. lenta, B. lutea, B. populifolia* and *B. nigra*), trees unknown in the King's Garden. In Canada, birches were called "Providence trees," perhaps because, like bamboos in China, they were used for everything. Boats could, of course, be made of them, and the thick bark of some species concealed a woody material both solid and flexible which was suitable for covering tents as protection against bad weather. The Indians used the sap of certain branches for medicinal purposes. Beginning in the Garden, birches soon spread out into the forests of France. Sarrazin also dispatched cypresses, acacias, maples, bilberries (*Vaccinium arboreum* and *pennsylvanicum*), and hundreds of flowers peculiar to the cold regions—liliums, arums (*Calla palustris*), and angelica (*Aralia nudicaulis* and *A. racemosa*).

These collections were to be completed by a most well informed man, Roland Michel de La Galissonnière. A sailor who was to become governor of Canada from 1747 to 1749, he attempted to form around him a scientific nucleus. He was himself interested in botany, but the local flora did not seem to arouse much enthusiasm in the others. La Galissonnière believed that the reason for this indifference was that Canada is primarily the land of the hunter. Its animals—beavers, marmots, and badgers—are fascinating. The ermines play games, the polar foxes are full of pranks; there is fishing under the ice, blood on the snow. But the marquis de La Galissonnière was not a man to be easily discouraged. Throughout his time as administrator, he struggled to supply the King's Garden with seeds, watching over the dispatches with jealous care. He also dreamed of vast botanical "exchanges" consisting of sending to France the great forest trees of Canada and planting in Canada fruit trees from French orchards.

La Galissonnière's name would have appeared in histories of botany only as a man of good will, however, if he had not one day, with infinite pains, sent to Paris a tree which was to fill all Europe with such enthusiasm that it became the most sought-after ornamental tree—the *Magnolia grandiflora*. The existence of the tree had already been reported. It is even possible that some magnolias had already been sent to France from Asia, since several species existed there. But those specimens were not the magnificent American grandiflora. Moreover, they must have been poorly transported, or perhaps merely described but never sent, since the first officially registered magnolia is definitely that of La Galissonnière. Like all plants at the time, it too arrived in Europe at Nantes, adapting there so well that Brittany, even today, is the region of the continent having the largest number of these magnolias. Later, the tree was named by Linnaeus to honor Dr. Pierre Magnol (1638–1715), a botanist who first thought of classifying plants by families and at one point was the physician of Louis XIV.

La Gallissonnière had found his *Magnolia grandiflora* in the south of Canada, a fact that caused much surprise in Paris, for botanists were convinced, from previous reports, that this tree grew only in hot countries. In the second half of the eighteenth century other species were discovered in warmer climates and some temperate regions of Latin America. But fossilized specimens have also been found in Greenland, Alaska, and northern Sweden.

After returning to France and to the navy, La Galissonnière died in 1756 with one last regret; he had not been able to bring back to France the horse chestnut (*Aesculus octandra*), which grows in Illinois. We also owe to him the discovery of the giant tulip tree (*Liriodendron tulipifera*), the flowers of which look like green tulips. This treasure trove of new plants was eventually classified by Linnaeus, thanks to his pupil Peter Kalm, who roamed through Canada with Dr. Gaulthier and sent seeds to his master.

The magnolia of La Galissonnière survived, chiefly because acclimatizations in the King's Garden had become, by this time, almost always successful. Collaboration between the three major botanical centers, Montpellier, Nantes, and Paris, produced spectacular results. Slowly but steadily the soil of France was enriched with new species sent by the missionaries, scientists like Tournefort, administrators, explorers, and sailors like Flacourt, la Salle, and La Galissonnière. There were so many new trees that Fagon, with the approval of Louis XIV and his ministers, had established what was to be the first true nursery. Its experimental nature was underscored by the fact that the enterprise was closely connected with the Crown. No action having to do with the nursery was permitted without the signature of the king. He even chose the site, an area situated between what today are the Champs-Elysées and the rue du Roule in Paris. There, under the watchful eye of Fagon and his gardeners, more than 20,000 new trees were grafted. These could be bought for a very moderate sum by anyone interested in them. The "Pepinière du Roule," as it was called, also contained an orangery for delicate plants and was directed by an exceptional botanist,

Sébastien Vaillant, a favorite collaborator of Fagon and also the man whose small coffee plant had been the inspiration for the Garden glass houses.

The Garden and the botanists also suffered some tragic losses. In 1705, one of Fagon's protégés, the young botanist Augustin Lippi, together with an entire diplomatic mission, was assassinated on the way to Abyssinia. Lippi was only twenty-seven years old. After landing in Cairo, the mission was blockaded somewhere along the Ethiopian border, and finally massacred. Augustin Lippi had been able, however, to send off a few messages before the disaster, and he had also dispatched some seeds to Fagon. The latter was convinced that Lippi would have been a great botanist, and there is much to confirm his view. Lippi was the first to be interested in algae, for example, which no one before him had seriously thought of including in the realm of botany. Before leaving for Egypt, he sent numerous algae specimens to the Garden from Marseilles along with a lengthy report. Danty d'Isnard, a professor at the King's Garden, sought to reconstitute Lippi's work about 1710, but it was too widely dispersed. At least 200 specimens sent back in his name had been classified in other herbaria. Later, Michel Adanson was to pay tribute to the perceptiveness of Lippi's observations on palms and fig trees, noting that his predecessor had been the first to call attention to the existence of the baobab (*Adansonia digitata*), which Adanson was to introduce into Europe. And as a posthumous tribute, Linnaeus dedicated to the murdered botanist a plant of the verbenaceae family, the lippia.

The massacre of the Abyssinia mission helps us understand the perpetual state of anxiety experienced by members of most of the scientific expeditions of the period. These terrible losses of men, time, and energy aroused considerable frustration. The tireless Fagon was the only one never to despair. Indeed, as time went by, he pressed on yet more urgently. The years 1712–15 were a period of reassessment and of preparing the future. The Garden superintendent demonstrated again his extraordinary talent as a discoverer of men. "Discovered" on the present occasion were the three Jussieu brothers, Antoine, Bernard, and Joseph.

The three were sons of a well-known Lyons pharmacist, Christophe de Jussieu, who made his name by publishing a treatise on the Greeks' famous theriac. This medicine was composed of about a hundred different plants; it had the power either to kill or to cure, depending on the addition or removal of certain ingredients. Like their father, the three brothers were lovers of medicine and botany. All three were to be appointed to positions in the King's Garden, and founded a veritable dynasty, since after them Antoine-Laurent, Bernard's nephew, and Adrien, Antoine-Laurent's son, were also to be famous. The first to come to Paris was Antoine, who was born in Lyons on July 6, 1686. When Fagon called upon him to accept the directorship of the Garden, Antoine de Jussieu was already a famous doctor and an excellent botany professor. In 1715, at the age of twenty-nine, he was elected to the Académie des Sciences. Watching him carefully, Fagon immediately recognized in Antoine what he would indeed prove to be, a powerful and enlightened scholar, of remarkable lucidity, able to foster, help articulate, and then synthesize a wide range of new ideas. Like Fagon, Antoine de Jussieu felt that the time for taking stock had come. First, he set about a long period of theoretical reclamation, reinvestigating the existing travel narratives, studies, and notes, himself publishing numerous lectures and addressing to the Académie des Sciences the first important study of coffee. At the same time, he carried out the first significant botanical "export," sending to Martinique coffee plants acclimatized in the Garden. By so doing, he provided the French overseas territories with a new source of wealth and secured coffee for France. Finally, with the help of the botanist-gardeners, still under the direction of Sébastien Vaillant, he checked all the Garden plantings.

Meantime, the curtain fell on an age: Louis XIV died in 1715,

Fagon in 1718. The king had reigned seventy-two years, and Fagon had been his Garden's superintendent for fifty-four years. The thirty-two-year-old Antoine de Jussieu immediately had to face the new overseers, Fagon's successors, men whose competence and enthusiasm left much to be desired. In 1722 he summoned his younger brother Bernard, then in Montpellier, to replace Sébastien Vaillant in the position of "lecturer on plants." Bernard, who was born in 1699, was a doctor, but he could not stand the sight of blood, which was a distinct disadvantage, especially in the eighteenth century. Therefore he accepted with alacrity his brother's invitation to become "lecturer" in the Garden. In addition, he worked in the laboratory, wrote numerous treatises for the Académie des Sciences, of which he became a member in 1725, and corresponded with scholars all over the world, in particular Linnaeus. His work of classification and acclimatization in the Garden and his steady support for the traveling botanists proved invaluable, as did his theoretical writings and his remarkable *Traité des vertus des plantes*. His career, in fact, closely resembled that of his brother.

But there was a third Jussieu, Joseph, a turbulent character, amazingly gifted—perhaps almost too gifted. His studies, marked by great powers of assimilation, were brilliant but disorganized. In addition to being a botanist, he later became a doctor and mathematician, even wishing at one point to be an engineer. Antoine and Bernard, wise and solid men sought after for their good judgment, were worried by their brother, who seemed unable to control his adventurous spirit. They did not yet know that an enthralling country would one day drive Joseph insane.

The Jussieu brothers: Antoine, Bernard, and Joseph.
(Paris, Bibliothèque Nationale. Photo by ERL)

Le comte de Buffon. (Paris, Bibliothèque Nationale. Photo by ERL)

[7]

Buffon *and* Jussieu

OUIS XV, the *Bien-Aimé*, was on the throne of France. The *Grand Siècle*, the century of Louis XIV, had been busy venturing new ideas, experimenting, trying to reorganize a world up to then too dominated by God to be rational. Now came the Enlightenment—a time to think, to confirm, and to gather into the same perspective all the recently acquired knowledge, and then go beyond. Throughout Europe philosophers, artists, and scholars were restlessly confronting their theories. The universe became understandable. What the great Linnaeus was to do for botany, Buffon, born the same year—1707—would try to achieve in natural history.

Georges Louis Leclerc, comte de Buffon, came from a noble family of Burgundy. When he entered the Académie des Sciences in 1733, at twenty-six, he became the central character in the scholarly world of Paris. A man of impressive stature and status, a wealthy landowner and ironmaster, he had studied law (to please his father, a counselor in the *Parlement*), medicine, mathematics, and physics; he was one of the translators and most fervent supporters of Newton. He was even a skillful swordsman, having been forced in 1728 to leave Angers, where he was studying medicine, because he had killed in a duel an officer with whom he had argued. This universal specialist, like many scientists at that time, was chiefly interested in zoology and botany. This latter interest led, in 1739, to his appointment as overseer above the director, Jussieu, of the King's Garden. He had just undertaken a series of studies of woods and forests. His appointment was a stroke of good fortune for the Garden, which, for the first time since Fagon, had a supervisor who was more than a mere administrator.

The year 1739 marked a turning point in Buffon's life. Until then, he had been a "budding genius lacking an object," as Sainte-Beuve was to put it. But now he would be able to devote himself exclusively to the study of natural history. He had a grand scheme—to write a complete history of the animals, plants, and the earth which bore them. He fulfilled this dream with his vast *Histoire naturelle*, but even so enormous a project did not occupy all Buffon's energies, or express, except in a synthesized form, the full range of his interests. We can better understand the breadth of his intelligence by considering some of his other writing: From 1734 to 1748, while compiling his great work, he presented to the Académie des Sciences not only his *Journal d'observations sur les aurores boréales* but also papers on the *Cause du strabisme*, the *Loi d'attraction*, the *Découverte de la liqueur séminale dans les femelles vivipares*, and the *Miroirs ardents pour brûler à grande distance*.

Buffon admired nature, but he also feared her disorder, which filled him with horror:

> Brute Nature is hideous and dying; I, and I alone, can render her pleasant and living. Let us drain these marshes, bring to life these stagnant waters, by making them flow, forming them into streams and canals, utilizing this active and devouring element, hitherto hidden from us, and which we have learnt to use by ourselves. Let us set fire to this useless growth, these old, half-decayed forests, then cut away what the fire has not consumed. Soon, in place of the reed and water lily, from which toads make their poison, we shall see the ranunculus and

the trefoil, sweet and healthful herbs. . . . A new Nature will be shaped by our hands.

The praise of ruins was not for Buffon. Such was the dual face of the eighteenth century, a strange blend of preromanticism and pre-industrialism.

The new supervisor substantially renovated the Garden and extended it. No sooner had he taken up his position than he improved and developed the trial grounds by cutting apart plots of land to make flower beds. Antoine de Jussieu took advantage of this activity to arrange the plants in a new order. Buffon added to the fossil collection and to the number of live animals in the Garden, for he had very definite ideas on these matters. He preferred animals to everything else, but, in a deeper sense, he believed that more could be learned about the vegetable world by understanding the animals which eat plants or which eat other herbivorous animals. The new supervisor strove therefore to develop the *Cabinet du Roi*, the future Muséum d'Histoire Naturelle. In 1739 this establishment contained only two rooms, in which were set up a herbarium, a pharmacy, various minerals, and animals either fossilized or in glass jars. To enlarge the Cabinet, Buffon unhesitatingly gave up his official apartments, and moved to the hotel Lebrun, in the rue des Fossés-Saint-Victor. Because the Garden also needed to be extended, land between it and the Seine was acquired, after long negotiations with the owners, the monks of the abbey of Saint-Victor. Teaching interested Buffon as well. He had a large amphitheater built. Soon it was frequented by many students attracted by the quality of the professors. The botany courses given by Antoine de Jussieu, Louis Lemonnier, and René Desfontaines were particularly brilliant.

From every point of view, this was a rich period for the Garden, one which lasted for half a century, until Buffon died in 1788. Such a period was, in fact, long overdue, for although the Garden was in a fine state in comparison to the early years, things had scarcely changed since Fagon's death in 1718. The slowdown was largely due to lack of credit. Buffon, for example, had learned with amazement from Antoine de Jussieu that he had been forced in recent years to pay out of his own pocket the cost of plant transportation, fertilizer, and even tools, without ever having been reimbursed. In 1729 Jussieu had had to bring two small cedar plants back from England in his hat, and to pay himself the cost of shipping 100 seedlings of the tree, given to him by an English friend. The two cedars from his hat are now enormous, still alive in the Garden. Buffon could obtain what others had been denied because he had made his mark in society and could move at ease through the corridors of power. This freedom was a considerable asset at a time when power was concentrated and absolute, when ministers, and only ministers (together with the king), decided upon scientific missions.

The current concern was to participate in the scientific movement of the day. Henceforth, although not completely dependent, the Garden and its chairs of botany were linked to the Académie des Sciences, itself at the center of the intellectual upheaval. Scientists, who had never seen such a flurry of exploration and experimentation, were also busy verifying what their great forerunners, such as Newton, had said. Newton had maintained, among other things, that the earth is perhaps not as round as men believe, that it must flatten out at the poles and bulge in the middle. This had to be checked. The Académie des Sciences decided to measure the length of the arc of the meridian of one degree in Paris; then the comte de Maurepas, navy minister, was to send the mathematician Pierre de Maupertuis, who had introduced Newton's ideas to the French, to do the same thing in Lapland. Finally, the Académie would compare the two figures. In order to be absolutely certain, they decided to ask others to carry out the same calculations under the equator. These "others" are those who concern us.

Buffon liked the Jussieu brothers and admired their work.

When an equatorial expedition was discussed, Buffon managed to arrange for Joseph de Jussieu to accompany the group as a botanist. The Académie could not miss the opportunity to learn more about South American flora, the atmosphere in the forests, the varieties of trees, and the composition and uses made of local woods. Wood was of particular concern to France because it was needed to restock the forests and build boats. The mission took shape: Apart from Jussieu, the prominent members of the group were the astronomer Louis Godin, the hydrographer and geometrician Pierre Bouguer, and the group's leader, Charles-Marie de La Condamine. La Condamine was one of the great adventurers of the eighteenth century. The son of an important family that had belonged to the immediate entourage of Louis XIV, Charles-Marie was born in Paris in 1701. Following the family tradition, he chose an army career and took part in the wars in Spain. But, like Descartes and Maupertuis, he was a scholar-soldier, and his brilliant studies in chemistry and geodesy won him election to the Académie des Sciences in 1730. In Paris he found scientific circles in a state of confusion because of Newton's theories. These immediately interested La Condamine. The Académie des Sciences designated him to measure the arc of the meridian under the equator. He even participated financially in the venture, putting the sum of 100,000 livres at the organizers' disposal. The adventure was to last ten years, and La Condamine was to return a sick and weakened man.

On May 16, 1735, in a state of euphoria, the mission set sail from La Rochelle. On June 22, after an uneventful crossing, the scientists reached Martinique, the first working stop. They climbed Mount Pelée, and Joseph de Jussieu sent his first batch of seeds to the Garden. These contained nothing very new. Father Plumier had already traveled the same road. They next reached San Domingo, where they remained for three months to explore every inch of the island. They sent back some new consignments of plants, but suffered from the first attacks of fevers which claimed the life of Jussieu's servant and later struck the entire expedition. Nevertheless, the mood remained optimistic, judging from letters sent by Joseph to Antoine. On August 11, 1735, he wrote: "This letter is just to let you know that we have reached San Domingo safely. I am wonderfully well, eagerly gathering for you everything botanical that might interest you. As I begin to familiarize myself with the subject, I no longer find it so strange. I am delighted to be able to polish up my knowledge here; I shall now be better fitted to profit from my researches in Peru."

They did go to Peru, but in stages, by way of Cartagena in Colombia, where Joseph botanized a little. They next stopped for two weeks in Porto Bello, but Jussieu spent the first week sick in bed with fever, for which he was bled. His troubles were beginning. Nonetheless, he managed to undertake some research, found some rare plants, visited the gold mines—even sending some nuggets to France—before proceeding to Panama. There the harvest of plants was so plentiful that to list them here would be tedious. Everything was sent off to the Garden; there were even, wrote Jussieu, "many other plants which I have described and had drawn, but whose seeds were not yet ripe. Still others I have only been able to preserve dried." We may note in passing that he gave a long, detailed description of the Panama purpura, which furnishes a most beautiful dye.

The group was in Guayaquil in April 1736. During 1737 and 1738 Jussieu did not write home. News reached France from La Condamine. Finally, on August 31, 1739, they reached Cuenca, in Ecuador, where the mission surgeon, Seniergues, lost his life in a riot. "My dearest brother," wrote Joseph, at last, "I was longing to send you an account, on returning from my trip to Hosca, of the most noteworthy things I had seen in this province of Peru. But I was prevented from doing so by the illness of Monsieur Seniergues, who has been badly wounded. Those of us who are French were almost killed in a mass uprising. Fortunately, the mob did not carry out its plan to exterminate us all." The journey did not seem to be going so well.

By "most noteworthy things," Jussieu was referring to the fact that he had, despite everything, fulfilled one of the chief aims of his mission—to discover and send to Paris as soon as possible the mysterious quinquina tree, which supplied the miraculous "Jesuits' powder" Louis XIV and Fagon wanted so much to obtain. Crossing the Andes from Ecuador to Peru, Jussieu had passed through forests of quinquina (*Cinchona officinalis*). The longed-for tree was soon growing in the Garden greenhouses. The bark of this tree, a member of the rubiaceae family, furnished quinine and cinchonine, two tonic substances that lowered fevers, something the Indians had always known, jealously guarding the secret until 1600. The precious powder had then crossed to Spain. It reached France in 1678 (Fagon used it successfully to cure Louis XIV of a bout of swamp fever). But the French had yet to find a source of the powder for themselves. Plumier had died just before setting off to hunt for it. Joseph de Jussieu had at last sent the precious plant to Paris, accompanied by some mémoires which were used by Linnaeus, years later, in formulating his description of the genus *Cinchona*.

Joseph de Jussieu in fact pulled off a double coup. He also sent back a shrub which had caught his attention because he noticed the Indians chewing its leaves with evident enjoyment. When questioned, they explained that the shrub was called the coca and that its leaves cured everything. Jussieu corrected their claim, explaining that it merely soothed pain. The coca clips (*Erythroxylon coca*) reached the Garden, were planted, and grew. It was not until 1865 that the pharmaceutical panoply was enriched with the new "strong analgesic," cocaine.

Meanwhile, Joseph de Jussieu felt strangely tired. Not ill, he insisted, just tired, and occasionally quite disoriented. When the feeling passed, he would set off again through the jungle, sometimes on horseback, sometimes on foot, always alone, protected by the Indians he did occasionally cure and who taught him all their knowledge about plants and animals. He was in a constant state of amazement. Peru disturbed him deeply.

La Condamine was not wasting his time either. Not only did his studies confirm Newton's hypotheses, but he also undertook the reputedly impossible task of tracing the first map of the Amazon river, proving the connection between the Amazon and the Orinoco, a link hitherto dismissed as an idle fancy. He traveled, of course, in jungles abounding in plants. And although botanical research was not La Condamine's main interest, he could not but think of Jussieu who, immobilized by a high fever, had stayed behind in Peru. He noted down for his friend everything that struck him as interesting. One day, he was impressed by the various uses to which a certain resin could be put: "When fresh," he wrote, "it can be molded into any shape. It is impervious to rain, but even more remarkable is its great elasticity. It can be made into bottles which do not break, or into boots and hollow balls, which flatten when pressed only to return to their original shape when the pressure is released." La Condamine learned that the resin came from a tree called the *cahuchu* (*Hevea brasiliensis*). This major discovery marked the beginning of an industrial and commercial revolution. The first rubber trees were sent to the Garden greenhouses.

La Condamine's proper mission had been over for some time, but he was now inventing tasks for himself no one had asked him to perform, apparently unwilling, or unable, to return to France. Was his reluctance the result of the splendor and mystery of this land, of the torrid, enveloping atmosphere of the jungles, or did it have to do with the lianas? Let us listen to La Condamine daydreaming: "The lianas climb and twist around trees and shrubs in their path. Having reached the branches, often at a great height, they let fall perpendicular runners, which bury themselves in the earth, take root, and grow up again, rising and falling by turns.

Other runners, carried obliquely by the wind or by a chance move-ment, often attach themselves to nearby trees, forming a tangle of hanging cords, in every direction, looking just like the rigging of a ship. Some are as thick or even thicker than a human arm; others choke the tree they surround, finally killing it in their embrace." La Condamine appeared to be succumbing to the same embrace.

When, in 1745, he finally returned home, ten years after he set out, he was forty-four years old, ravaged by fevers and half paralyzed. He lived on, however, with an exquisite creature at his side, a niece whom the pope permitted him to marry. With in-credible determination, he continued to write as long as he could, describing everything, the mighty river, the jungle, the rubber, of which he had discovered another variety (*Hevea guyanensis*) in Guiana; he even described curare, the uses of which he discovered in Amazonia. "This poison," he reported, "is an extract made over a fire from the juice of various plants, particularly of certain lianas. They tell me that curare contains over thirty kinds of herbs and roots." La Condamine also performed numerous experiments, de-stroying the legend that sugar was an antidote against curare.

His works were published in Paris as they were completed. La Condamine was still extremely active. He defended the ef-fectiveness of vaccination against smallpox and fought for the adop-tion of a universal measuring system. To him we owe the choice of the French *toise* as a legal measure, prefiguring the present metric system. Gradually his entire body was paralyzed and he became deaf. When he died in 1774, all the scholars of Europe, his corre-spondents, attended the funeral.

Joseph de Jussieu, meanwhile, was alone, for, their work finished, Bouguer and Godin had left, the former returning to France and the latter settling in Lima to teach mathematics. La Condamine had headed for the rivers, Séniergues had been killed, and two other scholars had become insane. But Jussieu lingered in Peru, unable to tear himself away. In each of his rare letters, he mentioned his imminent return, which did not take place until too late. Despite his absence, he was elected to the Académie des Sciences in January 1743 for his achievements: From him, the King's Garden received the Peruvian nasturtium (*Tropaeolum minus*), the most colorful and beautiful of all, and a hydrangea (*H. radiata*), a plant already mentioned by the missionaries in Asia and also found in North America. Joseph sent back orchids, declaring Central America to be the favorite habitat of these flowers. Their arrival in Paris aroused great excitement. Another important discovery was the cinnamon (*Canella alba*), called *quixos* by the natives. "This is a very tall tree," Joseph told Antoine, "the wood of which is most fine and suitable for carpentry. It has the additional advantage of being sweetly scented. The leaves are long, like those of our laurels, the flowers and fruit resemble those of the genus described by Plumier as *Borbonia*: we may appropriately name the tree *Borbonia peruviana*."

Many other packages were sent, and Joseph noted: "I am sending you the other species I was able to save from accidents and humidity. Those I gathered at Quito are fresh and may, if they reach Paris rapidly, bear fruit in the King's Garden, satisfying your taste for botanical curiosities." These plants were the "delicately scented heliotrope" (*Heliotropium peruvianum*), the fine *Magnolia sausurus*, a periwinkle (*Vinca*) from the Andes, and a plant the Indians called *ipecacuanha*, our ipeca (*Cephaelis ipecacuanha*).

Despite undoubted achievements, the arrangement was no longer working as it should. Jussieu was wandering from the cordillera of the Andes to Pacific ports. Soon forced to be self-sufficient, he practiced, in order to survive, all the trades he had more or less mastered, working, for example, as a doctor in Quito in 1743. Sometimes, his solitude seemed to bear upon him. In 1744, he wrote to Antoine: "I did not think, dear brother, that our journey to Peru would entail an absence of over ten years. I could bear such a separation when I was still receiving news of you, but for the last five years I have been deprived of this, my only blessing, and have fallen prey to the most profound melancholy. I have written to you several times and sent various packages of seeds, which I am convinced were lost, because the war interrupted the movement of trade. I hope to be more fortunate this time, and that the present consignment will shortly be followed by my arrival in France." Then Quito was ravaged by smallpox. Forbidding Joseph to leave the city, the governor ordered him to care for the sick. "Just when I wanted to leave," complained Joseph. He stayed on and became interested in the various epidemics devastating the city. He discovered, for example, that the dust from the volcano Cotapaxi created a severe illness in the Indians. But he continued to suffer from "deep melancholy," fevers, and vertigo. Whenever he was well, he left to explore new regions, always alone and unarmed. He visited the Cannelle valley, a difficult, hot, and exhausting region. He had to cross the mountains and climb the volcano Tunguragua, 16,690 feet high. At last, Jussieu received a letter from France and enough money to pay for his trip back home, as well as Godin's, who was still in Lima. Jussieu left Quito to rejoin Godin, and both put all their notes and specimens in good order. In October 1746, while they were making their last preparations, Peru was devastated by an earthquake. The port of Callao, from which they were to sail, was entirely destroyed by a gigantic tidal wave. The scientists had saved their lives and their belongings, but the viceroy ordered them to help the authorities. Jussieu tried to cure the wounded, Godin directed the reconstruction. It was only in 1748 that they decided again to leave. This time they were to cross South America and set sail from São Paulo. They crossed the Andes yet again. In Bolivia they visited Cuzco, which Godin did not know. Then, following the valley of the Urubamba, they reached Lake Titicaca, where Jussieu assembled a collection of aquatic birds. The two men finally arrived in La Paz in July 1749.

But Jussieu was caught again by his anxious and melancholy mood. One night he left to go back to Lake Titicaca. He had not warned Godin, who went back to Europe alone.

This pattern of behavior continued—a haphazard itinerary, conflicting projects, and fevers, endless fevers, which resisted treatment with quinquina. He kept collecting plants, animals, and observations, which he sent to Lima. He was persuaded that, some day, they would eventually reach France. Jussieu seemed unaware that time was passing him by. He seemed oblivious to the fact that he was growing old and was a sick man, slowly losing his mind. He felt he had one last thing to do: to visit the infamous mines of the Bolivian province of Xauregui. Jussieu was so horrified by the condition of the Indian miners he found there that he stayed in Potosi for four years, trying to soothe their miseries. Searching his memory for the lessons he had learned in the days when he wanted to be an engineer, Jussieu even had to construct and repair bridges, dikes, and roads. In 1755, however, the governor of Xauregui had to return to Europe, and he suggested that Joseph accompany him. But when Joseph reached Lima in an extremely weak state, it was to learn of a series of misfortunes. His mother had died, as had several of his brothers, in particular, Antoine, who had been his chief correspondent and for whom Joseph felt a kind of filial piety. With so many bridges now burned, why should he bother to go back across the sea? He also learned that a great many of the plants he had sent had been lost. He was stricken with fatigue and depression. The governor had to leave, but Jussieu no longer had the heart to accompany him. He remained on in Lima, where, more than ever, he was to bury himself in a lonely existence, scraping a living from his medicine, renewing his study of mathematics and mechanically setting his notes in order. He wrote to Bernard de Jussieu, in March 1762. "I was trusting, dear brother, that I would be able to send you . . . a big batch of plant seeds. But my bad health, in particular the dizzy spells that frequently overcome me, have prevented me from going out of town to botanize." Jussieu's dizzy spells were perhaps no more than the continuation of the vertigo he had experienced ever since his first day in this disturbed and disturbing land, with its jungles, orchids, and the fever-bearing volcanos.

He carried on for another eight years, exploring mountains and jungles once more, but matters went from bad to worse. His new notes were burned, the old ones ruined by damp. Perhaps in Paris he might be able to gather everything together again. But in Paris they were growing worried—Jussieu must be brought back before it was too late. His brother Bernard alerted mutual friends and sent emissaries to Joseph. One of these messengers eventually got through, and, on July 10, 1771, thirty-six years after his departure, Joseph de Jussieu returned home. But the man who came back was but a shadow, his body worn, his head forever empty. Surrounded and cared for by his family, honored as a martyr to science, he lived on for eight years, writing nothing, never regaining his reason or even going out of doors. When he eventually died, on April 11, 1779, it was said that he seemed to have vanished in a dream. After Jussieu's return to France, his family and friends realized that his crates had been left behind. They asked for them, but everyone in Quito or Lima had forgotten that the sick, rather insane old man had been a great scientist and that his crates contained treasures accumulated through years of incredible exploration. The crates were lost. Nothing was left of the works of Joseph de Jussieu.

The supervisor of the King's Garden, Buffon, who had sent Joseph de Jussieu to Peru, was more fortunate in his other endeavors. Influenced by him, taught by him, or "discovered" by him, a number of bold men set sail for the remotest regions of the globe, to send back treasures of observations, animals of all species, and plants to the Académie, the Cabinet, and, of course, the Garden, which was rapidly becoming an important scientific center.

Buffon greeting a visiting monarch. (Photo by Roger-Viollet)

Watercolor of an orchid painted by Joseph de Jussieu.
(Paris, Bibliothèque Centrale du Muséum National d'Histoire Naturelle)

[8]

In and around Africa

THE Court in the eighteenth century was still dreaming of China and regretting the Jesuit disaster. Painters, such as Oudry, Boucher, and Pillement, did what they could to satisfy this nostalgia, but they merely prolonged it. The workshops copied Chinese porcelains, Aubusson and Gobelin tapestries were covered with Chinese motifs, and the word *Chinoiserie* was on everyone's lips. Then, a young twenty-two-year-old botanist, Pierre Poivre, asked to be sent on a mission to China. His request created a great stir. Everyone wondered who the young man was and what exactly his mission entailed. Lemonnier, the Jussieu brothers, and Buffon all met him. After listening to and understanding his ideas, they decided to accept his offer.

Pierre Poivre, Peter Pepper, in English, seemed to bear a predestined name, but it was pure chance. He had nothing to do with the discovery of pepper (*Piper nigrum*), which Theophrastus had mentioned long ago as the berry of an Arabian vine. But the coincidence did perhaps play some part in his life, since he was to devote many years and all his energies to spice trees, "*épiceries*" trees as they were called then in French, supplying Europe with a plethora of new perfumes and tastes.

Pierre Poivre was born on August 23, 1719, in Lyons, his parents' hometown. They were the heirs of a family of silk manufacturers who had lived in the city for 300 years. The town and its way of life were both eminently practical, based on trade. Only the river Rhone provided an invitation for daydreaming. Pierre's parents intended him for the priesthood, and he began to study theology with the missionaries of Saint-Joseph. There he also studied ancient languages, natural science, and mathematics. He learned so well and so fast that his masters decided to send him to their seminary for overseas missions in Paris.

Could this first trip, made entirely overland, have set Pierre dreaming of the sea? Contemporary interest must also have played a part, for the sea had by now provided young people with heroes. Landlubbers longed for a life on the ocean during this great period of pirates, buccaneers and privateers. After making rapid progress and finishing his studies at seventeen, Pierre Poivre was still too young to take orders. Moreover, he was not quite sure that he wanted to become a priest. It was not that he lacked faith, but he feared a priest's life might be too sedentary. Since he had plenty of time before him, his superiors suggested that he go to China and learn Chinese.

The botanical world in Paris then was at its most brilliant. Linnaeus, the great Swedish scholar, had just come to the capital to meet Buffon, the Jussieu brothers, and Louis XV. His arrival heralded a revolution, for he spoke endlessly about the sexuality of plants and the mysteries of their fertilization processes. Beguiled, the king requested some private conversations and even presented Linnaeus with some seeds which the scholar took back with him to Sweden. Botanical gardens were multiplying: One had just been created at the Trianon, headed by Bernard de Jussieu, and two others, at Auteuil and Marly. Everything planted in these gardens came from the King's Garden. Everybody had heard of and was talking about these events. The world was dazzled by flowers, and the number of students in Buffon's lecture hall doubled. Pierre

Poivre was soon fired by the same enthusiasm. For four years he studied botany, and he learned enough to impress his teachers. The idea of China was still on his mind, and soon he was on board a ship belonging to the Compagnie des Indes, en route for Mauritius, then known as Île de France. He embarked from there, in 1741, immediately for China.

In Canton he introduced himself as a scientist. But was he not also something of a missionary? And missionaries, since the expulsion of the Jesuits, were not entirely welcome. In short, for reasons that are not quite clear, what Pierre Poivre first discovered in China was its prisons. Even here, he took advantage of his situation to learn Chinese. He learned it so well (later he was to learn Annamite and Malay) that he succeeded in charming the prison governor, who boasted of the studious nature and high moral qualities of his prisoner. When the story reached the ears of the governor of Canton, he asked to meet the young prisoner, who emerged from the palace a free man, bearing letters which would permit him to circulate freely in the province.

Pierre Poivre had no interest in seeking revenge for past troubles. There to see China, he set to work to do just that. As he did so, he reflected on the unrealistic expectations that had led Rome to seek to evangelize this land of long traditions. As a result, Europeans were now shunned. Even scholars like Poivre were unwelcome, and the missionaries had been forced to take refuge in nearby countries—Tonkin, Siam, and Cochin-China. He went to visit these missionaries and brought back from Cochin-China numerous observations on woods, star aniseed (*Illicium anisatum*), pepper, betel (*Piper betle*), pineapple, (*Ananas cochin-chinensis*), coconuts (most probably *Cocos nucifera*), and their oil, indigo (*Indigofera tinctoria*), mulberries (*Morus alba*, introduced in Europe in 1494), and the raising of silkworms, and mangroves (*Rhizophora*). He also wrote a study of rice. Then he returned to China, but soon left again, feeling as if he had somehow failed to complete his task.

Pierre Poivre's tribulations had just started. On January 16, 1745, he left China aboard the *Dauphin*, a fine 750-ton ship armed with ten cannon. When they were attacked by the English, Poivre joined the crew in defending the vessel, but the English won the fight and took the French sailors prisoner. Pierre Poivre's right hand was almost torn off in the combat. His notes and the plant specimens he was carrying with him all went down with the boat. Still not rebelling against his fate, Poivre learned to write with his left hand. The English allowed him to land at Batavia for an amputation; the entire arm had to be removed. But in Batavia he did discover what from now on would be his life's inspiration—spice trees. The magical perfumes and dazzling beauty of these plants were enhanced by their considerable commercial value. At that time the Dutch completely controlled the spice trade; the islands were guarded by soldiers and attempted thefts were punished by death.

After five months in Batavia, Poivre knew that he had to win a spice empire for France. He wanted to bring the king nutmeg and clove trees; for clove trees, "when fully mature, often bear several *quintaux* (hundredweight) of cloves a year." This proved impossible, but by the time he left, Poivre had formed a plan that he knew would interest the French. He set sail from Malaysia on board a French brigantine, the *Favori*; adverse currents and high seas caused the vessel to founder off Siam. Unperturbed, Poivre boarded another boat, bound for Pondicherry, where there was a branch of the French East India Company. From there, he felt sure he would be able to return to France. More tribulations awaited him. He did indeed find a boat, but the vessel had to return by way of Mauritius, to pick up La Bourdonnais, the governor, who was returning to Paris. But they ran afoul of the English. To gain time, Poivre changed to a Dutch boat, which was later stopped by a privateer from Saint-Malo, which was itself captured by the English. Poivre ended up a prisoner in Guernsey. After much bargaining, he was freed and returned to his country.

Given such experiences, other men might have relaxed for a while, but Pierre Poivre, having seen and heard so much about which he was longing to talk, determined not to have lost his arm in vain. He presented his plan to the Compagnie des Indes, to Lemonnier, Buffon, and Louis XV. He wanted to turn Mauritius and Reunion, then called Bourbon Island, both French possessions with the necessary climatic conditions, into a kind of enormous extension of the King's Garden. Both places would be used to grow exotic ornamental plants and for plantations of nutmeg and clove trees. This new source of supply would break the Dutch monopoly of fine spices, which they sold all over the world at exorbitant prices. His hearers listened attentively, for Poivre spoke in economic terms. Spices began to seem more and more attractive: The trade in cloves and nutmegs alone, Poivre asserted, brought the Dutch a huge annual profit (of 50 million pounds). "I finally learned," he concluded, "that the continued ownership of these profitable spices, the basis of Dutch power in the Indies, was made possible chiefly by the ignorance and cowardice of the other European trading nations. In order to win part of the inexhaustible wealth which the Dutch had hoarded secretly in this remote corner of the earth, one needed only to be aware of the situation and have the courage to seek to share it with them." The solution was to obtain a few small specimens, grow them in the good earth of Mauritius, then sell them at competitive prices.

Pierre Poivre explained to Buffon and Lemonnier his wishes to make Mauritius a private botanical garden as well, and to collect there all the plants of the Indian Ocean. Because it was both new and solidly argued, the proposal attracted the masters of the Garden. Poivre was, in fact, offering to raise on the island plants grown from seeds gathered at the right moment. Such a procedure would render the King's Garden independent of sea transportation, the precariousness of which was all too well known to Poivre. It would also avoid the constant problem of plants rotting in ships' holds, after which they were fit neither for experiment nor acclimatization. If, in the future, his consignments were lost, Poivre would immediately be able to send a new batch from Mauritius. Moreover, he promised to include with each shipment notes about the problems of acclimatization; these notes would deal with exposure, soil composition, the flowering period, and so on. The proposal was accepted immediately.

The *Jardin des Pamplemousses*, the Grapefruit Garden, was born; some of the finest flowers were to be nurtured there. Scarcely was he back in the Indian Ocean region in 1749 than Poivre sent the King's Garden six new species of scented acacias, balsams (*Impatiens grandiflora*), amaranths (*Amaranthus viridis*), pomegranates (*Punica granatum*), irises, grapefruits (*Citrus paradisi*), doubled-flowered lilies never before seen in Europe, the scented, so-called Chinese lilac, the citron tree (*Citrus medica*), the most sumptuous hydrangeas, and a fine ornamental tree, the wild nutmeg (*Monodora myristica*), which Bernard de Jussieu immediately acclimatized. Other boats brought back the cinnamon (*Cinnamomum zeylanicum*), the lichee (*Litchi chinensis*), the sago palm (*Metroxylon laeve*), the latania, the soapberry tree (*Gymnocladus chinensis*), and various camellias. These new arrivals were greeted enthusiastically.

Pierre Poivre, back again in what he called the "torrid zone," was now duly commissioned by the authorities. But what he had not foreseen, despite his speech on the need to be bold, was that he himself would have to become a freebooter and a thief in order to obtain nutmegs and cloves. These were hard truths to face. In addition to the Dutch monopoly, there was a black market in spices, controlled by murderous groups of rival pirates. Pierre Poivre had a difficult task ahead of him. From Mauritius he made a trip to Annam, where he had friends and could buy seeds and cuttings. He sent Lemonnier another cinnamon tree (*Ci. cassia*), the aloes wood (*Aquilaria agallocha*), or eaglewood, more pepper plants, more camellias and hydrangeas, and more rice, with notes on the last-named's various properties and cultivation sites. He also included a strange tree, the Madagascar clove (*Ravensara aromatica*), praising the beauty of its flowers and fruit, which was described as "a green nut whose shell and inside have a taste similar to that of the clove." Flacourt had already mentioned seeing the tree in Madagascar.

Poivre still had no real clove tree (*Eugenia caryophyllus*), and no nutmeg (*Myristica fragrans*). Setting off in an old camouflaged skiff, he landed by night in Manilla, foiled the sentries, and managed to escape with nineteen nutmeg plants under his cape. He was pursued. Some plants were lost, others stolen, and yet others died on board ship. When he finally reached Mauritius, only five remained, but these few prospered. For the clove trees he had to go to Timor, where, from an intermediary, he bought twenty cuttings for 14,000 piastres. The plants were never delivered. His patience exhausted, Poivre organized another raid. He was almost killed, but he obtained his clove plants. He was well aware of the nutmeg's medicinal and other properties, such as its hallucinogenic powers similar to those of the ravensara. He pointed this out to his Paris colleagues so that they could investigate the matter further in the laboratory.

The Mauritius botanical garden had existed before Pierre Poivre's day, but in an embryonic form, having been set up in 1736 by the governor, La Bourdonnais. Neglected after his departure, it had been vaguely restored by the general administrator, Etienne Le Juge, and baptized *Jardin de Montplaisir*, Mount Pleasant Garden, only to be abandoned a second time. When Poivre had arrived, full of hopes and plans, things started to move fast. Anything he did not have at hand, he stole, sailing aboard small disguised skiffs, sneaking in at night through lines of Spanish, Portuguese, English, and Dutch vessels. He took amazing risks, but he obtained remarkable results. The Jardin des Pamplemousses, as he named it, was accorded the title of King's Garden, just like its Paris counterpart. What indeed could be more royal than this site? It is adorned with three hundred mango trees, with jackfruit trees (*Artocarpus integrifolia*), avocado (a local variety of *Persea gratissima*), bamboos (*Bambusa vulgaris*), acacias with flowers like birds' feathers, sour oranges (*Citrus aurantium*), bloodwood trees (*Haematoxylum*), a source for black and violet dyes, several species of cacao trees (*Theobroma cacao*), many varieties of coffee bushes, including those of Malabar and Madagascar, and, among the flowers, with the rarest of sensitive plants, althaeas, sweet peas (*Lathyrus odoratus*), and Bengal roses (*Rosa chinensis semperflorens*).

The granting of the royal title to the garden was also, surely, a way of thanking the man who was so effectively safeguarding both the acclimatization in situ and the transportation of his plants. Paris witnessed the arrival of some new nasturtiums (*Tropaeolum*), a giant "geranium" called the stork's bill (*Erodium alfilaria*), new gladioli, doubled-flowered pomegranates, white rocket (*Hesperis*), seven types of aloe, a superb local calla lily, ebonies (*Diospyros ebenum*) from four Indian countries, orange trees from China and Bengal, mignonettes (*Reseda*), and tuberoses (*Polianthes tuberosa*) from Goa.

Pierre Poivre, however, was tired—happy, no doubt, among his spices, but tired, or, rather, disillusioned. He had not done everything he had hoped to do. For instance, having read Etienne de Flacourt's writings, he had dreamed of compiling an inventory of the flora of Madagascar. He had visited the island, where he had seen enough to convince him that this place was in truth the botanists' paradise, but he had been in a hurry, rushing to catch a boat. He had brought nothing back from there, and could never return. He was tormented by the same feeling of a missed opportunity that had plagued him when he left China.

There were other reasons for his fatigue. Not the harsh law of the sea, which he had always accepted, but what he curiously referred to as "sailors' intrigues." These poisoned his relationships with men who, like him, navigators, naturalists, and Frenchmen, should have helped him in these far-flung regions. Instead, they acted only as rivals jealous of his successes. Certain climates seem to undermine even the strongest natures. In 1757, at thirty-eight, Poivre returned to France and his estate of La Freta in the Puy-de-Dôme, to concentrate on agriculture, marriage, and then his three children.

But the story was not yet over. In 1767 he was again asked to go to Mauritius, returning this time with the title of general intendant. Young Madame Poivre was the object of the galantries of the charming, smooth-talking Bernardin de Saint-Pierre, to whose advances she remained indifferent. Bernardin de Saint-Pierre was later to make of Madame Poivre the well-known heroine Virginie, in his *Paul et Virginie*, published in 1787. Pierre Poivre returned to France for the last time in 1772. He remained at La Freta until his death on January 6, 1786.

In 1768, a year after his nomination to the post of intendant of Mauritius, Poivre sent for his nephew, Pierre Sonnerat, to help him. He needed someone to keep up to date the immense mass of observations required for the reports on the spice-tree plantations and on the consignments sent to the King's Garden. Pierre Sonnerat was twenty when he landed in Mauritius. After successful studies in Lyons, he now had Pierre Poivre as botany teacher, and the Indian Ocean islands and the entire continent of Asia in which to conduct field work. Before long he would be well equipped to cope with his new task. Shortly after his arrival, the opportunity arose to go to the Seychelles, those islands to the north of Mauritius and Reunion, then said to be deserted. Almost nothing was known about these islands, which did not even all appear on maps. Terrifying stories were told, however, of giant crocodiles, prehistoric creatures said to frequent the island regions. Sonnerat said nothing about crocodiles, because he was far too intrigued by quite another marvel.

He landed on Praslin Island, one of the smallest of the Seychelles, and found it deserted indeed but covered with huge palm trees, the biggest he had ever seen. "On this island is found the palm tree which produces that very sought-after fruit, hitherto known as the sea coconut, the Solomon coconut, or the Maldive coconut (*Lodoica sechellarum*). The Island of Praslin, or Palm Trees Island, is, up to now, the only place where the tree producing this nut has been found." Sonnerat had reason to be proud of his first expedition.

The existence of this enormous fruit, a giant double nut that a small child could not get his arms round, was already known to Sonnerat and Poivre. Indians trading in these fruits sometimes sailed by way of Mauritius. No one knew where or from whom the Indians obtained the nuts, which were sold in the East for astronomical sums because of their reputed aphrodisiac powers. "Because this fruit was somewhat rare," wrote Sonnerat,

> its shape strange, and origin unknown, people attributed to it important properties, and invented stories about its origins. They claimed, and still believe, not only in the Indies, but all over Asia, that the nut of the sea coconut has all the properties we attribute, perhaps exaggeratedly, to theriac; and that its shell is a sure antidote against all kinds of poison. The great lords of Hindustan still purchase the fruit at high prices. They make the shell into cups, which they decorate with gold and diamonds, never drinking from anything else. They firmly believe that poison, which they greatly fear because they use it so often themselves, can never harm them, no matter how virulent it may be, if their drink has been poured into and purified by these healthful vessels.

When they were transplanted to Mauritius, the giant palms grew well, and, from there, they soon reached the King's Garden, where botanists could analyze the "medical" properties of sea coconuts, which weigh up to 50 pounds and require ten years to mature. Because of their suggestive shape, they were also called *Lodoica callypige*.

Pierre Sonnerat had a good boat and a capable crew. He continued his tour of the Seychelles, and landed next on the island of Mahé, the largest in the archipelago. He stayed only long enough for the ship to be supplied with fresh provisions, but he did have time to notice a thriving banana market. The winds forced him to set sail again, to his next ports of call, the Philippines and the Moluccas, or Spice Islands. There he carried out an extensive study of the breadfruit tree (*Artocarpus altilis*), so called because the flesh of its fruit, when cooked, tastes like bread. This fruit, he thought, could be the ultimate weapon against hunger.

With the coconut worth its weight in gold, the bananas, and the breadfruit tree, Sonnerat's research all seemed directed toward the useful, as if he were seeking to make botany commercially profitable. Such was indeed the case, and for quite simple reasons. Law's bankruptcy had ruined France and the royal treasury was empty. Under such circumstances, particularly at the period we have now reached, nations tended to turn to agriculture. Gone was the age of the vast forests with which Colbert had hoped to cover France. Men now had to eat and produce cereals, preferably corn, or anything that could be substituted for cereals.

But for all that, the age of flowers had not vanished. Louis XV loved them too much, as did his subjects, now so fond of gardens that they could not imagine life without at least a few flowerbeds. Nor had research ceased. It was watched over by Buffon, Lemonnier, and Bernard de Jussieu, and protected by the king. After Trianon, Marly, and Auteuil, two other botanical gardens were founded at Bellevue and Saint-Germain. The great sea excursions continued, usually with a botanist on board. Nothing, then, was completely lost—people were still buying orchids in Paris.

Pierre Sonnerat continued to travel from island to island. From the Moluccas he went to India. We may cite for its humor his discovery of the digitate-flowered cavalam (*Sterculia foetida*), also called "dung tree" because "its flowers," wrote Sonnerat, "smell of human excrement." His next stop was Madagascar. Wherever he went, he was drawn first to palm trees, those strange plants of unexpected beauty. He took cuttings everywhere, sending them first to the Jardin des Pamplemousses, then on to the King's Garden, where, under glass, they adapted to a new existence as ornamental trees. The palm tree soon became for Europeans the epitome of exoticism.

Methodically, Sonnerat undertook to trace the "geographic line" of this capricious tree, a line which did not always coincide with that of the 111°F of heat theoretically necessary for its growth. The tree could in fact withstand much lower temperatures, if the air, the ground, and its company were appropriate. One also had to specify which palm one was talking about, for there are almost a thousand different species. Some supplied dates, others coconuts, and the trunks of several (e.g., *Dracaena draco*) gave off a red resin called Dragon's Blood, to which many virtues were ascribed. Other varieties produced a kind of edible cabbage, and yet others wax. Even today, we know little about the palm tree, particularly about its fertilization processes, so complex that some have called it hermaphroditic. Pierre Sonnerat wrote important pages on the history of this tree.

After his 1774 trip to the Indian Ocean islands and India, Sonnerat went to Malaya and China. He returned by the same route, spending two years on the Coromandel coast before touching the islands and the Cape. He arrived in France in 1781 with an exceptional collection: 300 birds, 50 quadrupeds, butterflies, fishes, insects, reptiles, woods, and a "considerable herbarium." In 1789 he published *Voyage aux Indes Orientales et à la Chine*, a work destined to become and remain a classic.

Another tree was fascinating the king's gardeners then. It was very small when it arrived, but the accompanying notes were explicit: Do not be misled by the plant's present size. In this crate is no ordinary tree, but a baobab, a giant unequaled in the vegetable kingdom. Adanson had sent it from Senegal.

Michel Adanson, "the African," was born in Aix-en-Provence on April 7, 1723. He began to observe plants from the age of nine, small mosses which he grew on his windowsill. His family intended him for the priesthood, and at an early age he was appointed to a small canonry in Champeaux-en-Brie. His aspirations evidently did not lie in this direction, however, for he abandoned his post in 1745 when he was twenty-two. He devoted himself thereafter to the study of science and followed with special attention the botany courses at the Garden in Paris. Antoine and, particularly, Bernard de Jussieu became interested in the young former priest who made such rapid progress that, had he wished, he would have been admitted to the group of Garden scientists. But Adanson was naturally independent. His longing for freedom, heightened by a kind of misanthropy, soon proved incompatible with the pursuit of a normal scientific career. Well aware of this problem, and eager to make his name in the history of natural science, Adanson decided to travel. Given his state of mind, it was better to go where no one had been before, a consideration that determined his decision to leave for Senegal. Bernard de Jussieu was ready to help by finding for him a position as a clerk in the trading post which the Compagnie des Indes maintained there for its ships.

Africa in the eighteenth century was not crowded. There were few Frenchmen there. In fact, Europeans could happily have managed without this continent, which blocked the way to the Indies. The Portuguese had been the first to circumnavigate Africa, Bartolomeu de Novais by accident and Vasco da Gama on purpose. The Portuguese soon moved from the coast to the inland region, where there was an opening for trade. They did not regret their decision, for Africa proved to be a land of rich resources in the form of human labor. The inhabitants had little to sell and were hardly in a position to buy. They had only their capacity for hard work, a strength that was to be cruelly exploited. The age of worldwide slave trading began, a trade particularly necessary to South America, which had been somewhat depopulated by endless massacres. Sorely in need of laborers, the Portugese Empire shipped three or four million African workers to the mines of Brazil. The English, better organized than the Portuguese, also began a slave trade. France, with no vast empire to supply, limited itself to a small amount of slave trading on the coast of Senegal.

Adanson reached one of these trading posts, now Saint-Louis, on April 24, 1749, after a trying journey he vowed never to repeat. The climate there was difficult for someone fresh from Paris. Whites were said not to last long in Senegal. The whites there were all company employees, not particularly interested in natural science, and they did not establish any close relationships with the newcomer. Adanson was unperturbed, glad of the opportunity to devote himself to his studies. Senegal contained a few scattered company offices, in Podor, for example, on the Senegal river. Covered with thick bush or forest, everywhere else was inhabited by thieving nomads, elephants, and wild animals. In the river lived hippopotami, manatees, and crocodiles. This land was not always pleasant, but Adanson had to explore it.

It was on the banks of the river that he found the first specimens of baobabs, thick-set giants sixty feet high, with trunks a hundred feet in circumference. His guides told him that the tree was sacred, that its bark was used to make talismans and to treat dysentery and fevers, and that the leaves, when dried and powdered, could be made into effective medicines against abdominal and stomach ailments. Every native carried some of the powder with him. Every part of this beneficent tree was useful; its leaves and flowers served to make oil and soap, and its fruit provided a feast for the monkeys. Bats and night birds lived high up in its branches.

Adanson was especially intrigued by the thickness of the

trunks, made up of superimposed layers. Scratching and scraping at the bark with infinite care, he found on one baobab an inscription carved there by the English, exactly 200 years old, and covered by 200 layers of wood. Adanson pushed through to the center of the trunk and thus ascertained the age of that specimen. It was 6,000 years old. His amazement and delight are understandable, for no one had ever heard of anything like the baobab. His enthusiasm was soon shared by the scholars in the King's Garden and throughout the learned world. Linnaeus called the tree *Adansonia*, a name which Adanson always refused to use but which is still the scientific one for the baobab.

There were so many new plants in this nearly unknown world that Adanson filled entire crates with seeds, along with lengthy notes. A typical letter would be the following: "I am also enclosing the leaves of a species of Lawsonia (*L. inermis*), which the Omolofer Negroes call *Foudenn*. The powder is used by the local girls to beautify their nails. When mixed with enough water to give it a pastelike consistency, and applied to the nails for six hours, it painlessly turns the nails a rich, dark vermilion. The dye lasts for almost six months. When I tested the mixture on my toe-nails, it did not disappear for five months, by which time the nail had completely grown out." History does not relate whether the heads of the Garden conducted a similar experiment; probably, they took Adanson's word. *Foudenn* is in fact henna.

Adanson also found bergamot (*Citrus bergamia*), the citron (*Citrus medica*), the curcuma, a new pepper (perhaps the Guinea pepper, *Xylopia aethiopica*, or the Achantis pepper, *Piper guineense*), the vetiver (*Andropogon muricatus*), noting that it was a natural insecticide, and the citronella (*Andropogon nardus*). He shipped plants supplying gum and wax, some of them new species, others species already known and brought in by caravans in the sixteenth century. Adanson also sent back the mandarin orange (*Citrus nobilis*) and other fruit trees whose African names he kept and which it took the Europeans some time to identify. Many of these plants were from Asia, of course, probably carried to Africa by traders.

Finally, to the great satisfaction of Buffon, who was always interested in different woods, Adanson dispatched descriptions and samples of what were called "wood" trees, those suitable for carpentry and cabinetmaking, such as an araucaria, the so-called African larch, the silk cotton tree (*Ceiba pentandra*), which gives kapok, the ebony (*Diospyros ebenum*), and the touloucauna. Flowers were not forgotten either, the first from black Africa—local zinnias, new ageratums, convolvuli, and ipomoea; local myosotis, giant hibiscus, a new veronica, stocks (*Matthiola*), petunias, aloes, and a frangipani (*Plumeria*). Adanson also sent off an important study of dye plants, particularly indigo, which he had observed at length. It is worth noting that few of these plants are native to Africa. Zinnias, for example, come from the Americas; indigo is essentially Asiatic. These often appeared as such to the botanists because they had seen similar plants in the King's Garden.

Curiously enough, there were few adventures to report. Adanson experienced some health problems and a certain amount of friction with the employees of the Compagnie des Indes, but this first investigation of this new world did not produce any noteworthy incidents. Adanson was pleased, for he was not looking for adventure for adventure's sake. He soon fell in love with Senegal and its inhabitants and he became a lifelong adversary of slavery. He learned the various languages spoken by the Senegalese and, back in France, created a scandal by suggesting that a phonetic interpretation of French should be drafted for African use.

He had returned to France and a warm welcome on January 4, 1754, after four years and four months in Africa. He and Buffon concluded ten years of negotiations for the purchase of his collections by the King's Garden, in 1765, with the payment of a pension. Adanson was named king's botanist and assigned an official lodging in the Grand Trianon. In 1757 he published the first volume of his *Histoire naturelle du Sénégal*. In 1758 the king appointed him book censor, a position he held until the Revolution. In 1759 he described to the Académie des Sciences, of which he was a member, the plan of his most important work, *Les Familles des plantes*, which appeared four years later and earned him admission to the Royal Society of London. He planned a trip to Guiana which he had to abandon for lack of money. In 1766 Catherine II, empress of Russia, offered him a post as professor at the Saint Petersburg Academy of Natural Science, but Adanson refused.

He decided instead to compile a universal encyclopedia that would include all the natural sciences. This project was to run to sixty volumes with 40,000 plates and innumerable appendixes. The plan, however, was coolly received by the Académie, so Adanson, disappointed, moved into semiretirement, gradually breaking all his ties. He even separated from his wife and daughter, on the pretext that family life took too much time away from his work.

He did work, struggling to realize his dream of an encyclopedia. He owned a small garden where he grew more than a hundred species of mulberry and lived in a few rooms littered with sheets of writing. Scorning everything his predecessors and contemporaries had done, he rejected Linnaeus' nomenclature, substituting for it names borrowed from African languages. This futile effort isolated him even more. "When he retired to his cottage on rue Chantereine," wrote one of his last friends, "and had organized a small garden room for those plants he could still grow, he would spend the whole day there to observe them, sitting on the floor and dragging himself from one to another on his crossed legs."

Adanson lived to the age of eighty-three. When he died on August 3, 1806, the *Encyclopédie* of d'Alembert and Diderot and Buffon's *Histoire naturelle* had appeared, and most of Adanson's literary projects were never realized. But he had opened up a new part of Africa to French botany, and one of the largest of trees still bears his name, as does *Adansonia*, the review published by the Paris Muséum d'Histoire Naturelle.

Poivre worked on the east side of Africa, Adanson on its west coast. A third door to the vast continent was opened in the second half of the eighteenth century with the systematic study of Egypt's flora by Charles Sonnini de Manoncourt. In 1768 Buffon had become interested in this young prodigy. Sonnini was born in 1751, the son of a counselor of Stanislaus Leszczynski, sometime king of Poland, father-in-law of Louis XV, and, through the king's gift, sovereign of Lorraine and Bar. Sonnini was educated by the Jesuits of Pont-à-Mousson and had been granted the rank of *docteur en philosophie* at the age of sixteen. When he arrived in Paris, he was invited out, lavishly entertained, and met Buffon. Sonnini showed the utmost consideration for the master, with whom he enjoyed friendly conversations, and he soon became interested in natural sciences. But the young philosopher had to return to Lorraine, where his father wanted him to become a judge, a dismal prospect to the son, whose restless nature soon led him to choose a military career. This eventually led him in 1772 to Guiana, where he made good friends among the Creole privateers. The colonial administration, which rarely had a doctor of philosophy and a budding naturalist available, suggested that he draw up a list of the country's resources. Sonnini welcomed the suggestion enthusiastically. Exploring jungles and virgin forests, he opened up ground suitable for planting sugar cane, breadfruit trees, coffee, and other spice plants which he immediately sent for to enrich Cayenne. When he was promoted to lieutenant, he was called to Paris, where a small act of thoughtfulness won him admission to the entire scholarly world. While traveling in Guiana, he had remembered Buffon and brought back to him a collection of rare birds for his museum. Buffon, delighted, invited Sonnini to spend a winter in Montbard to help him put in order his studies on ornithology.

Realizing that Sonnini was a remarkable young man, Buffon

suggested that he go on a botanical mission to Egypt. Although he accepted with enthusiasm, Sonnini pointed out that he was not, in fact, a botanist. Buffon agreed, but he felt that Sonnini's extraordinary report on the possibilities for developing Guiana, his powers of observation, already well known in scientific circles, and the accuracy and liveliness of the letters he sent to the King's Garden all made him as qualified as would years of experience. And so Sonnini left for Egypt, from where he regularly wrote letters which were popular. When in Rosetta, for instance, he sent back the following report:

Among the useful plants in the Rosetta area I noticed the prickly pear (*Cactus opuntia*), the fruits of which the local inhabitants eat. Among trees I noticed the sesban shrub (*Sesbania aegyptica*), an acacia with yellow flowers and a sweet smell. I also saw the fig tree (*Ficus*), which has most attractive green foliage and spreading branches shading a wide area.

I later saw several fields, covered with the kind of large millet the Egyptians call Durra, a plant much cultivated here, yielding plentiful crops.

The large African marigold (*Tagetes erecta*) displayed its lovely yellow flowers among the plants in some gardens. . . . In the streets of Rosetta were sold stalks of fenugreek (*Trigonella foenumgraecum*), a plant used for forage and for medicinal purposes. Throughout the winter months, yellowing bunches of bananas decorate the Rosetta gardens. This species is the one whose fruit is known in the islands of America and in the colony of Cayenne. It is less insipid and pleasanter in taste than the common banana (*Musa*).

In addition to these exotic trees, I recognized another species I had seen in America, one pleasing both to the taste and the smell, namely, the Annona (*Annona muricata*). When transplanted into the gardens here, it reaches the height of a medium-sized tree.

In the shade of the orchards, they cultivate various plants, the roots of which are moistened by water running in all directions along small trenches. One sees many mallows, called *hobeze* (*Malva mauri-*

tiana). It is cooked with meat and is one of the most frequently used herbs in the kitchens of Lower Egypt.

Buffon was delighted to read such reports. From Embaba, a small village on the western bank of the Nile, Sonnini wrote again: "The fertile plains surrounding Embaba to the west are enriched by various kinds of cultivation. The inhabitants produce in particular a species, or, rather, a variety of lupin (*Lupinus termis*), much eaten in Egypt. Its seeds are cooked with water and salt, stripped of their hard, thick skin, then eaten."

Sonnini regularly described the food he ate—water melons, melons, and green dates. He also discussed at length hemp, or hashish (*Cannabis sativa*), much used by the Egyptians: "This kind of destruction of the power of thought, this slumber of the soul, bears no resemblance to the drunkenness occasioned by wine or strong liquors, and our language has no words to express it. The Arabs call this voluptuous abandon, a kind of delightful stupor, *kayf*." Such remarks probably did not offend Buffon, a man of the Age of Enlightenment. Sonnini followed with a description of the Doum palm (*Hyphaene thebaica*): "In the countryside are appearing the first shoots of a kind of fan-shaped palm peculiar to the upper region of Egypt. The Doum, as it is called, produces fruit twice a year. The fruits are rounded and slightly elongated, the size of an orange, but irregular in shape." And, further on, he noted: "On the dry and almost sterile plains of the same parts of Upper Egypt, there frequently grows the true acacia (*A. arabica*), which gives off gum arabic from its trunk and branches."

Sonnini mentioned everything—olive trees which did not supply enough oil anymore, mulberry trees and the silk trade, and the carob (*Ceratonia siliqua*), "a less precious tree, bearing fruit which is used in a special trade. The carob is common in other lands having a mild climate." The seeds were used as a standard of weight which eventually became the carat we use today.

He spent two years observing and describing, with great accuracy, the nature and customs of upper and lower Egypt, together with the area's resources, flora, and fauna, particularly its birds. Afterwards, Sonnini devoted himself more and more to zoology, publishing such classic works as his *Histoire Naturelle des Reptiles*. But it is, in a way, because of Sonnini's botanical exploration of Egypt that France was better prepared scientifically to undertake her later Bonapartian conquest of the land of the pyramids.

Chinese scene from a screen by Pillement. (Photo by Janet Goldwater)

Linnaeus explaining botany to Louis XV. (Photo by Roger-Viollet)

[9]

The Enlightenment and the Garden

THE increasing harvest of plants sent by the botanist-travelers to the King's Garden was safely grown, propagated, and even hybridized. Acclimatization, one of the main tasks of the gardeners, made important progress in the second half of the eighteenth century under the direction of one of Buffon's assistants, André Thouin. He was born in 1747 and he entered Buffon's life late. This slight disadvantage was rapidly overcome by the amazing precocity of the young man who was named gardener-in-chief of the King's Garden when he was seventeen years old. In a way, André Thouin was a Garden product, for he was the son of Jean-André Thouin, the chief gardener from 1745 to 1764, to whose position André succeeded. The succession happened neither by chance nor by inheritance. Young André had early demonstrated the most remarkable abilities, as, for example, when he accompanied Bernard de Jussieu in his botanizing expeditions.

Thouin immediately understood what Buffon wanted: With him he supervised the enlarging of the Garden, leveled the plots, and constructed terraces and walls, all in record time. He planted fruit trees, created a garden for economical plants, entered the Société royale d'Agriculture (and, later, the Académie des Sciences), before which he presented papers on rhubarb (*Rheum rha-*

99

ponticum), on the perennial flax (*Linum perenne*) of Siberia, and on Chinese hemp (*Cannabis sativa*). Working tirelessly, he gradually became the center of a group of correspondents who were interested in the whole world and who sought to circulate various plants from one continent to another.

Thouin had no equal as an acclimatizer. He had observed that the chief obstacle to the germination of plants from hot countries lay in the difference between the times the sap rises in the plants. By controlling the moment of growth, he accustomed plants to the effects of the Parisian climate, managing in this way to acclimatize the marvel of Peru (*Mirabilis longiflora*) and the dahlia, which was sent from Mexico to Madrid, and thence to Thouin, who favored it as a substitute for potato; he also acclimatized some exotic myrtles (*Melaleuca* and *Metrosideros*), species which now flower much later than when they were first imported from Australia and New Zealand. He also induced hydrangeas, daturas, and banisterias (*B. tetraptera*), to start their cycle in the spring while he encouraged the autumn growth of chrysanthemums, robinias, a red-flowered chestnut (*Aesculus x carnea*), the silver linden (*Tilia tomentosa*), an oak with sweet acorns, the Laricio pine (*Pinus nigra calabrica*), and so on.

Every year he sent out from the King's Garden more than 80,000 packets of seeds, 20,000 of which were sent to various monarchs and 12,000 to the French colonies for which he procured, among other things, the Tahitian sugar cane (*Saccharum officinarum*). He succeeded in implanting the bread tree (*Artocarpus integrifolia*), into America.

Buffon, Thouin, the Jussieu brothers, and the botanist-travelers were fortunate in that plants were exceedingly in fashion. "Another source of satisfaction to me is to learn that Natural History, especially Botany, has become popular with those in power at court," wrote Michel Adanson to Bernard de Jussieu in 1751. The great "favorite," Mme de Pompadour, who supported artists and writers, also adored flowers and instilled the same passion in her royal lover Louis XV. Flowers appeared everywhere: Dr. Louis Lemonnier, doctor-in-ordinary to the king, and also a naturalist and professor of botany in the Garden, worked zealously to help Mme de Pompadour decorate her garden at the Ermitage. Lemonnier ordered from the Garden various kinds of jasmine, myrtles, oleanders, pomegranates, lilacs, rhododendrons, and sensitive plants. The king was so happy in this garden that he asked Lemonnier to teach him the rudiments of botany, an art now at a particularly fortunate moment of its history because the fashion for English gardens was beginning to soften the excessive stiffness of the classical design.

France suffered many setbacks in the eighteenth century. Mainly because of negligence, the Indian empire assembled by Dupleix was lost to the English and closed to French travelers for many years. The duc de Choiseul, the influential minister of Foreign Affairs and of War (1758–70), decided to do away with the overpowerful Jesuits. Their establishments were closed, their property confiscated, and, in 1767, they were driven out of all the countries under Bourbon rule. The teaching of ancient languages, especially of Latin, was hard hit by their departure. The Enlightment, nonetheless, was at its apogee. It was encouraged by Louis XV, who was delighted by all forms of progress. In botany many works appeared that updated European knowledge of plant medicine, vegetable life, techniques of acclimatization, and the like. The most important theory, by far, was then that of Carl von Linnaeus. The Swedish scholar had finished formulating his classification system, which was adopted unanimously by all the botanists of his day and is still in force. With more difficulty, he had brought scholars to accept the existence of vegetable sexuality. He had revealed the complicated rites of fertilization, hermaphroditism, and self-pollination. These theories had already brought about a revolution, but Linnaeus then advanced yet another idea. He had believed in what was then called "fixism," the idea that there existed on earth "as many different

species as God had created forms at the beginning of the universe."
This plausible notion was now out of date since, as Linnaeus well
knew, everything he had experienced in the natural world suggested
the exact opposite of this theory. Now he began to ask whether "the
milieu may not have diversified the species, that is to say, one flower
might become quite other, evolving and changing itself so that in
another setting it looks completely different."

Although he was convinced of the correctness of Linnaeus'
theories, Buffon did not change his own position. He had no desire
to be involved in the coming quarrels. Only years later Jean Bap-
tiste de Lamarck formulated the new "transformism," which
opened the door to the theory of Darwin.

In an age when so many new ideas flourished and old concepts
were systematically verified, when men were trying to tie together
their new discoveries, it was natural enough that explorers would
do the same. The time had come for vast scientific expeditions
around the world, in which scholars would play a prominent role.
The first man to lead such an expedition in the eighteenth century
was Louis-Antoine de Bougainville, and he was accompanied by the
well-known botanist Joseph Philibert Commerson.

Where did Louis-Antoine de Bougainville, who was born in
Paris in 1729 in a family of lawyers, get his taste for the sea? His
father wanted him to inherit his notary's office, but Bougainville
showed his independence by deciding to become a barrister. Then,
to occupy his spare time, he devoted himself to natural science,
which was normal enough for anyone who was twenty in 1749.
This must have broadened his interests considerably because eventually he chose adventure and became a soldier. It was glorious to
make a career in the service of such a sovereign as Louis XV, who
had just declared, "We make war as a king, not a merchant."
France was then struggling to increase its power in India and was
fighting in Canada.

Bougainville left for Canada on April 3, 1756, to join the marquis de Montcalm's forces as a *commandant* in the Dragoons. Canada was eventually lost to England, but Bougainville was promoted to the rank of captain of the king's vessels because he had distinguished himself by his courage in the defense of that overseas French province. "I can assure you," Montcalm wrote of Bougainville, "that he has indeed a military mind." In France the young man found his king and country under attack from the best minds of the day—Voltaire, Diderot, and others were violently denouncing all forms of expansionism. Voltaire was particularly opposed to the Canadian adventure, writing that no one should fight for "quelques arpents de neige" (a few acres of snow). This republic of letters, inside the kingdom, was anathema to Bougainville, who longed for but one thing: to go far, far away, forgetting Canada and the fact that Montcalm, his general, had never received the aid he had requested. Bougainville had become something of a "lost soldier."

In 1764 he was charged with leading to the Falkland Islands (then called the Îles Malouines) a group of French-Canadian families who had refused to live under English rule. With two boats, the *Sphinx* and the *Aigle*, they made an easy trip, arriving without difficulty in the islands, which are near the Strait of Magellan. They were forced to stop for fresh provisions, and in the course of a stay in South America, Bougainville sent off to Lemonnier, for the Garden, the ingredients for making *maté*, a popular drink in Paraguay. *Maté* was brewed from the leaf of the *Ilex paraguarensis*, which resembled an orange-tree leaf but tasted of mallows. When dried and powdered, it could be made into a kind of tea which was drunk at any time of the day and which was there called "Jesuits' tea."

Bougainville made the trip to and from the Falkland Islands several times because there was virtually nothing on these treeless, rocky moorlands. Everything had to be made or imported if the settlers were to remain. In 1766 the Malouines had to be handed

over to the Spanish for political reasons. Bougainville, who disap-
proved of the change and who had brought life to those desolate
lands, had to hand them over himself to their new occupants. This
did not soften his bitterness. The king's order, however, also
provided prestigious compensation for the sacrifices it demanded:
Bougainville was officially ordered to return by way of the South,
and to make there all possible useful discoveries. This was to be the
first round-the-world journey by a Frenchman, preceding Cook's
by a few years. Before leaving, Bougainville was to return to Nantes
to take delivery of a new vessel, the *Boudeuse*.

At this point, in 1766, Louis-Antoine de Bougainville met
Joseph Philibert Commerson, a doctor, naturalist, and botanist,
who also disliked Voltaire's ideas. The two men were almost the
same age, Commerson being two years older. Both were lawyers'
sons, and they shared the same likes and dislikes. Commerson,
a correspondent of Linnaeus, was as famous in the world of botany
as Bougainville was in that of the navy. He had excellent qualifications, having been trained at Montpellier. Among his numerous
skills, natural science was always his chief passion. But Commerson
was not an easy man: Secretive and passionate, he was also extremely
ambitious. He was eager to prove to himself that nothing was impossible for him. As a result, he had numerous accidents. When
gathering flowers, he would fall into a ravine or be carried away by
a raging torrent. In Montpellier they had to bar him from the
Garden, because he had been stealing from it in order to enrich his
private collection. Ignoring the ban, he scaled the Garden walls at
night. In this way, he had, at an early age, become the owner of
one of the finest plant collections in France.

Botanizing, for Commerson, bore no resemblance to Rousseau's solitary wanderings. He prospected, traveling widely in the
Cévennes, the Pyrenees, Provence, and the Swiss and French Alps.
Near Geneva he met Voltaire, who invited him to become his secretary. Commerson, however, found in Voltaire not "Promethean

fire" but "the look of a crook." He painted the following portrait of Voltaire, which would delight Bougainville: "A damned soul, a shade wandering on the banks of the Styx, whom one had to follow everywhere in order to write down even his nighttime terrors." Commerson had no intention of living in such company, so he returned home to the Ain region, to Chatillon-sur-Doubs, where he was born. There he set up a small botanical garden in which he observed the plants he had collected. He wrote constantly and his notes earned him a reputation that soon led to his being invited to Paris by Bernard de Jussieu. A tall, handsome man with dark, shining eyes, he surprised and charmed Paris. Commerson was described as having "an ardent and impetuous nature, violent and extreme in everything, from gambling to love, in his hatreds as in his friendships." He is said to have disliked the pleasures of society, and to have been indifferent to hardship and difficulty. He did marry, however, and when his wife died giving birth to their son, he was plunged into despair. He was not a man to make a new life for himself. Why not, then, go round the world with Bougainville, as the duc de Praslin, minister of the navy, and the king himself suggested? Commerson had met Louis XV, who had been so pleased with him that he had immediately conferred upon him the rare title of botanist and naturalist to the king.

The idea of a world trip was tempting. But Commerson, who suffered from pleurisy, could not leave his four-year-old son, now being raised by a maternal uncle, without first doing everything possible to secure his future in case the worst should happen. He took all possible precautions with his will, selling one property and planning out the returns on his rented farms on his other estates. Then he could prepare his mission. The king asked for plans and reports about what he expected to find. His essays (which he claimed to have written only to serve as a reply "to those of my relatives and friends who, if something happened to me, might be so foolish as to ask 'What was he doing in the Austral lands?'") were so remarkable that Praslin had them copied to serve as models to all those who, in the future, might lead or take part in similar expeditions. Commerson, who was also a doctor, seemed to have an extraordinary presentiment of death. He closed off his future even as he moved forward into it. He wrote a *Martyrologe de la botanique* which was discovered after his death. His friend Lalande, the work's eventual publisher, simply added to it a chapter on Joseph Philibert Commerson.

In December 1766 the *Boudeuse*, commanded by Bougainville, set sail in the direction of the Strait of Magellan. She joined the other ship of the mission, the *Etoile*, in Rio de Janeiro. The sea trip was so long that Bougainville and "the eminent naturalist, Monsieur Philibert Commerson," as the ship's log calls him, had time to talk. They came to know and respect one another even more. The sea journey, though tedious, was less subject to surprise attacks than travel overland. The South American coasts, where they occasionally had to stop, were, however, particularly dangerous, for they might fall into the hands of the Spanish or Portuguese. In Rio de Janeiro the almoner of the *Etoile* was killed by the Portuguese. Sometimes, when there were no problems, Commerson could disembark, as in Montevideo, from where he wrote to his brother: "Winter is about to begin here, just as summer starts in Europe. I have not forgotten to gather plenty of plants, birds, and fish, and would like to miss nothing, but how? A hunting or fishing trip or a walk here places me in the situation of Midas, beneath whose fingers everything turned to gold. This country is the most beautiful in the world. Oranges, bananas, and pineapples follow one after the other. The trees never lose their greenery."

On board, the crew grew impatient, wondering when they would reach the strait. When they realized that they lay even farther away than Patagonia, some of the men deserted. Bougainville coped with these problems easily, while Commerson, unmoved, remained in his cabin, bent over his plants. When they landed, Bougainville wrote, "Monsieur de Commerson spent the days gathering plants. He had to overcome all kinds of obstacles, but this rugged terrain had for him the attraction of novelty. The Strait of Magellan enriched his books with a great number of unknown and interesting plants."

Sailing from island to island, they eventually reached Tahiti, the New Cythera, as Bougainville, dazzled by the Tahitian women's beauty, named it. Bougainville proceeded to draft maps and to make notes on the climate, countryside, inhabitants, customs, and agriculture. The last was naturally rich because coconuts, bananas, breadfruit trees, and sugar cane grew plentifully. In exchange, reported Bougainville, "we sowed corn, barley, oats, rice, maize, onions, and vegetable seeds of all sorts for them. We have reason to believe that the plantings will be well cared for, because these people seemed to us to like farming."

Commerson noticed a tree which grew everywhere in Tahiti but which had never, so far as he knew, been mentioned. It was "a splendid plant, with broad flowers of a sumptuous purple, the ornament of many houses." Why should he not use it to pay tribute to his friend? Some time later, the first bougainvilleas reached the King's Garden after which they were sent all over the world.

In 1768 they reached Mauritius. Bougainville continued alone because Commerson was too tired to follow him. He was not a good sea traveler, and, besides, he enjoyed the Jardin des Pamplemousses and the company of Pierre Poivre, who welcomed him, ensured his rest, and worked with him as soon as he was well enough. In 1770 Louis XV sent word that he wanted a study of Madagascar, where he proposed to extend the French plantation. Commerson, still tired, set off again, for six months of happiness on the Great Island. "Madagascar," he wrote Lemonnier, "really is the naturalist's promised land. There, nature seems to have retired to a private sanctuary to work with other models than those she has followed elsewhere. The most unusual and wonderful forms are found at every step."

In 1771 Commerson was back in Mauritius, his beloved herbaria now enriched with Malagasy treasures. But fate was closing in. Pierre Poivre left for France, but Commerson could not go with him because he fell ill again. He was in despair, all the more so because Poivre's successor, Maillard, had no appreciation for natural science. Gathering his last strength, Commerson decided that he too would go home. He organized the dispatch of specimens brought back from Madagascar, together with all his herbaria and manuscripts, working with the same care and gloomy energy as when he made his testamentary arrangements in France. He had no illusions about the future. He had committed the sin of pride by refusing to communicate to anyone the key to his work, hoping, once back in France, to work alone on his discoveries and classification system until the end. Now he knew that he was going to die, and he tried to save his work despite his illness. "I am more bogged down than ever," he wrote to a friend. "I can't take a step now without being in great pain." He died of pleurisy or pneumonia, and perhaps of grief and loneliness, at the age of forty-six, in March 1773. A week later, because news of his death had not yet reached Paris, he was elected to the Académie des Sciences.

A year after Commerson's death, Bernard de Jussieu received thirty-four enormous crates of vegetable material. Later he declared that he had never received so many cases from one traveler. Bernard's nephew, Antoine Laurent de Jussieu, who became *démonstrateur* at the Garden in 1790, estimated the number of species sent by Commerson at five thousand, of which three thousand were new. Not one could be acclimatized. Commerson's herbaria were integrated with the huge general herbarium of the Garden. Bernard de Jussieu and those who had followed his travels or with whom he had corresponded tried to gather together his work and letters. He was too well known and liked to be soon forgotten. But all that remains of him is the bougainvillea, which does not even bear his name, and some plants he had brought to the Jardin des Pample-mousses from Polynesia, like the "Cythera," or Otaheite, apple (*Spondias cytherea*).

Why did Commerson fail? A major factor was undoubtedly the excessive secrecy with which he surrounded his work. This secretiveness was caused by a suspiciousness of character that ap-proached paranoia. "In your hands," he wrote Lemonnier, "I feel my secrets to be safe. But permit me to observe that the republic of letters is like a beehive: it contains many fat, lazy drones who live at the expense of the busy worker bees. I have often been attacked by these hungry, treacherous creatures." The chief problem, however, was perhaps that Commerson, who almost went round the world, was in fact more a book man. Herbaria were always his true passion. His pressed flower collection contained no fewer than two hundred gigantic volumes, and he spent almost all of his energy composing them. He worked feverishly to assemble these huge books of plants: The flowers had to be neither too open nor too young, so that the petals would be firmly attached. Then came the painstaking work of arranging the flowers, stems, and leaves, and waiting for them to dry before delicately transferring the plant to the final paper. Commerson performed many of these tasks at night, absorbed in solitary amazement. His friend Lalande describes him as "without rivals, friends or helpers, spending entire weeks, both day and night, going without sleep, determined to examine and arrange the treasures his expeditions had provided or correspondents had sup-plied. He was seen to spit blood after several weeks of such work. He was often found, his light still on, long after daybreak, not having noticed the sun come up." After his death, his work, scattered and anonymous, disappeared into the dark recesses of museums.

But Commerson's experiences provide a good illustration of the duality and potential dilemmas inherent then in botanical endeavor. Botany was at times an adventure of movement, at times one of sight: The botanist had to find and bring things back, but he also had to observe and understand. Commerson's fate gave rise to endless discussions between Buffon, closer to his written works, and Lemonnier, Thouin, and Bernard de Jussieu, the great plant distributors. All agreed on one point: A number of risks would be avoided if, instead of entrusting plants to ordinary boats, they could make use of vessels specially equipped for transporting plants, with competent gardeners on board.

Commerson was not forgotten by his friend Bougainville. After returning in 1769 by way of New Guinea, the Cape of Good Hope, Ascension Island, and the Cape Verde Islands, a journey that took two years and four months, he lived until 1811 and became one of Napoleon's senators. He published his *Voyage autour du monde* in 1771 and remained in the scientific world. He had a busy political and military career and held a brilliant salon in Paris. He always encouraged traveling botanists, to whom he never ceased to recommend the working methods of his friend Commerson.

The monarch who had sent Bougainville around the world, Louis XV, died in 1774. He spent the last months of his life roaming the woods, declaring that trees were his true passion. His ill-fated grandson, Louis XVI, inherited his love for plants, one of the few interests he shared with Marie Antoinette. Louis XVI purchased a huge estate at Rambouillet to serve both as experimental center and nursery. Guidance came from the directors of the Garden, and Thouin sent to Rambouillet each of his newly acclimatized plants, to be cared for there by skilled gardeners. When hunting and lock-making left him some spare time, the monarch enjoyed going to Rambouillet to practice plant grafting.

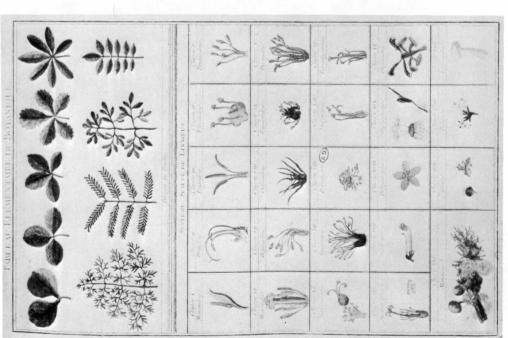

*Examples of plant classification: left, the method of Linnaeus; right,
that of Tournefort. (Photo by J.-L. Charmet)*

A draftsman recording the observations of a scientific mission.
(Paris, Bibliothèque Nationale. Photo by ERL)

[10]

The Revolution and the Garden

T HE trip of Bougainville around the world and the publication of his *Voyages* aroused considerable interest. But his navigation had been hindered by inaccurate maps, and he did not explore all the lands he discovered. On the other hand, Captain James Cook, who had set sail for his first trip around the world a year before Bougainville returned, had thoroughly described what he had observed. The English had acquired, through him, a monopoly of knowledge about the faraway islands.

Louis XVI, therefore, in conjunction with his ministers and the Académie des Sciences, ordered a new voyage. The men who would lead it were to verify the latest discoveries and, if possible, make more. Such a venture called for a first-class captain, full scientific apparatus, and a team of scholars able to cover all disciplines. The lack of this kind of variety had been another flaw in Bougainville's enterprise.

There was no problem in choosing a leader for the expedition. The obvious man for the job was Jean-François de Galaup, comte de La Pérouse. The best sailor of his day, he had recently distinguished himself against the English in Hudson Bay. His scientific staff, gathered in accordance with Buffon's suggestions, covered a wide range of disciplines. It included an engineer, Monneron; a geographer, Bernizet; a surgeon, Rollin; an astronomer, Lepaute-Dargelet; a physician, Lamanon; two draftsmen, Duché de Vancy and Prévost (draftsmen in those days were very important; they played the role of modern photographers and cameramen); a clockmaker, Guery; a universal scholar especially strong in chemistry and geometry, Gaspard Monge; and, finally, two botanist-naturalists, Boissieu de la Martinière and Collignon. To the last-named Thouin entrusted a full dossier of the current state of botany. He indicated every plant known to the Garden, included notes on successful and unsuccessful acclimatizations, mentioned the plants which had been reported but not yet discovered, and finally gave directions to be pursued in their research.

Probably no expedition had ever been prepared with such care and organization. Aboard two brand-new ships, the *Astrolabe* and the *Boussole*, the group set sail from Brest on August 1, 1785. Thirteen days later they were in Madeira, and, on August 19, they reached Tenerife, where Monge disembarked to study climatic phenomena. This was, at least, his professed reason for leaving the group, but perhaps he was filled with a strange foreboding.

The journey was a fabulous one, around Cape Horn, then on to the enigmatic Easter Island and the Sandwich Islands, to the north of which La Pérouse discovered Necker Island. They reached the northwest coast of America, and then, by way of the Mariana Islands, they went on to Macao. After sailing beside the Philippines in the direction of Japan, they made another important discovery, that of the strait separating the islands of Hokkaido and Sakhalin which today bears the name La Pérouse. From there the expedition set sail for Kamchatka, then headed south for Samoa, continuing on to Australia and Botany Bay, where La Pérouse wrote his last letter on February 7, 1788.

Everything had not gone as smoothly as the above might suggest. Early on, Cape Horn had not been rounded without difficulty. The two boats had encountered terrible storms and had often been in considerable danger, usually because of their pilots' ignorance of safe landing places not indicated on the available maps. Worse, the commandant de Langle, captain of the *Astrolabe*, had been killed by the natives on Samoa.

La Pérouse had planned a long stay in Mauritius to allow the scholars to rest and set their notes in order. These notes must have been substantial, judging from the ships' itinerary. But we shall never know what treasures they contained, any more than we shall know the exact circumstances of the shipwreck, which happened off the Vanikoro atoll in 1788, and of its tragic consequences. There were no survivors. A few letters from La Pérouse, addressed to the Académie des Sciences, did survive, as did some packets of seeds Collignon sent to the Garden as the ship's stopovers and chance meetings allowed. Thouin analyzed these packages, but was not able to profit greatly from them. The most ambitious scientific undertaking of the ancien régime had foundered: Three years later the king who had ordered the mission was imprisoned; he was guillotined in 1793.

Suddenly it looked as if the wind had turned. Many other botanist-travelers in the late eighteenth century were to fail or to die tragically. The most tragic case was perhaps that of Dombey. Joseph Dombey, born in Mâcon in 1742, was the son of a confectioner who wanted his boy to study pharmacy. But Joseph had other ideas and changed from pharmacy to medicine. In 1765 he became a doctor in Montpellier, a title which meant less than nothing to him, for he had in the meantime conceived a passion for botany, the result of his acquaintance with Antoine Gouan, a friend of Linnaeus. One of Dombey's biographers states that because of this new interest "he spent several months gathering plants everywhere, compiling many herbaria which he would later discard with the utmost casualness." Another biographer states that "he was a born traveler." Dombey did indeed travel, in Switzerland, in Savoy (whose people, he later decided, "bore the closest resemblance to the Peruvians, because of their situation and gentleness, their frugality and industriousness"), on the Mediterranean coast, and in the Pyrenees. Despite his so-called casualness, he always returned with rare specimens.

He went to Paris in 1772 with a fine herbarium to show Bernard de Jussieu. Jussieu introduced him to Lemonnier, Thouin, and Jean Jacques Rousseau, but it was André Thouin who was to become Dombey's closest friend. Here is Thouin's description of the thirty-year-old Dombey: "His was an attractive physique, slender but well-proportioned. He had large dark eyes and thick eyebrows of the same color. His dark skin gave him an African appearance. Although good-tempered, trusting, and open, Dombey was sometimes very melancholy. He was polite without being vapid, proud without being haughty. He particularly liked helping the unfortunate, even his enemies." The latter, as will be seen, were to show him no gratitude for his kindness.

In Paris the masters of the Garden soon viewed him as one of their own. He gathered plants around Paris with Rousseau, but he preferred to spend his time with Thouin. He never tired of watching his friend's miracles of acclimatization and grafting. Imitating the Chinese, Thouin had just begun to practice grafting, to the despair of the Garden scribes, who could not decide where to classify all his beautiful new hybrids. While Dombey studied plants, Louis XVI's minister of finances, Turgot, asked the Académie des Sciences to organize a mission to Peru. A former pupil of the Garden schools, where he studied chemistry, Turgot continued to show great interest in that venture. He wanted to find out the extent of Peru's resources of cinnamon, saltpeter, and platinum (the last substance was much in demand then for the manufacture of precision instruments). When Bernard de Jussieu was consulted, he naturally mentioned Joseph Dombey.

The problem was that the Spaniards controlled Peru. Nothing could be undertaken without authorization from the king of Spain. Permission was granted, but only after long negotiations and on the condition that Dombey was to be accompanied wherever he went by two Spanish scholars who were to share all his observations. These scholarly intrigues, carried on behind closed doors, forced Dombey to wait patiently in Madrid for almost a year. He learned Spanish and, finally, was told that his future companions would be Hippolito Ruiz and Joseph Pavon. These two men had been charged, unbeknownst to Dombey, with overseeing all of the expedition.

They set off on October 28, 1777, and on April 9, 1778, reached their destination. Dombey was fascinated by the world he discovered when they landed, and he was at once interested by the cacti. Until that time the French had failed to appreciate the cactus. They were familiar with certain small species that grew in glass houses and came from one or other of the French colonies, but nothing which they knew was particularly striking to the imagination. All this was to change with the arrival of the much grander specimens sent by Dombey. Their flowers, fruits, and shapes amazed everyone, but, more than anything else, people were puzzled about the exact nature of the cacti. What were they really? Were they even plants at all? Could we be certain that they were not in fact animals? When he studied the cactus, Dombey felt that he was crossing a mysterious frontier, even though he was convinced that they were, in fact, vegetable. Thouin agreed with him, having received in the King's Garden the first cereus cacti, which can grow to a meter in height. One variety of cactus, the "old man cactus" (*Cephalocereus senilis*), grew a sort of wool that, according to Dombey, "is shorter but of the same color as that of the Peruvian sheep," the llama. Other varieties soon followed. Public success was immediate. Everyone soon boasted of having a cactus.

In addition to cacti, Dombey shipped home some new convolvuli, the finest species of begonia of which Thouin made a sought-after decorative plant, amaryllis, lemon-scented verbena (*Lippia citriodora*), an arum growing in the Andes, the monkey-puzzle tree (*Araucaria*), and twenty different species of bignonia, which soon replaced the common bignonia known in the West. He also sent giant sages (*Salvia splendens*), and a tree, the "Guayaquil cedar," from Ecuador, the wood of which is completely impervious to water. He sent back herbs for dyeing, flowers from the Andes, and a stunning collection of new orchids, confirming Joseph de Jussieu's instinctive feeling that Peru was the favorite habitat of this ultimate flower. The orchids, as may be imagined, became the rage of Paris, and at the end of the eighteenth century there appeared a new type of collector, the wealthy orchid fancier.

Dombey was keenly interested in the medicinal or utilitarian aspects of the Peruvian flora. He collected Indian recipes for preparing potatoes and made extensive observations on a tea substitute, the *manglillo*, as well as on the *saya-saya*, a cure for angina, and the *michana*, whose roots Peruvian ladies used for cleaning their teeth. A single excursion around the capital enabled him to gather 300 plants, at least half of them unknown. He also carried out some geological digs.

Soon, however, his luck changed. Dombey's first consignment for the King's Garden was lost—but not for everybody. The boat carrying his crates was stopped by an English privateer, who sold the booty—quinine, cacao, copper, and tin—in the port of Lisbon to the Spanish government. Suspicion and fear clouded the relationship between the botanist and his two Spanish companions. More than anything else, however, Dombey found Peru itself unsettling. He could not adjust to the constant climatic variations: "When I went out for a walk, the climate was like that of Paris in the spring," he noted one day. But, soon after, came the desert sands, too hot for men and beasts, followed by the Andes cordillera, with its exhausting changes of atmosphere, and then the heavy, suffocating forest, which seemed to swallow up any non-Indians who ventured into it. The humidity rotted the travelers' provisions: "I went plant collecting, ax in hand," relates Dombey, "and for every plant I had to hack down a tree. The heat and stagnant air, combined with the strenuous work of cutting down trees, drained all my strength. It was impossible to regain our energy, because we lacked food." Moreover, in most areas the Indians were rebelling against the Spanish administration, so much so that Spain began to believe that Peru was going to slip permanently from its control. The mission itself was sometimes attacked by natives.

But Peru was the source of yet another malady, one that would acutely afflict Dombey: the fascination with vanished civilizations. At every step he was disturbed, worried, attracted both by the dead cities, the ruins of the Inca world, and by their replacements, like the Jesuit cathedral constructed on the foundations of the Temple of the Serpent. The sense of death that emanated from the deserted cities disconcerted Dombey as much as the scurvy he had just contracted. His difficulties were compounded by a feeling of fatigue which, as La Condamine and Jussieu had discovered, seemed more acute here than in Europe. After he had been in Peru for four years, Dombey longed to leave. To ward off the sense of absence and distance, he gave the plants he discovered the names of his Parisian friends. To Thouin he dedicated the *Thouinia multifida*, to Daubenton the *Aubentonia*, and to Lamarck the *Marckea*.

He did not, however, go home right away. First he explored Chile, the land of myrrh, laurels, and quicksilver mines. When he returned to Peru, he prepared his crates, into which he piled guavas, passionflowers, coffee plants, lianas, and specimens of the tuberous geranium that grows in the shade of rocks. He wanted to travel with his crates, to be present when they were opened, and to witness Thouin's delight. But he was tired and ill, often losing his balance and falling. Finally he set sail for Cadiz and arrived on February 28, 1785. But the seventy-eight crates destined for the King's Garden were confiscated because the Spanish government demanded that their contents be shared, a procedure that took three months. When Dombey reached Paris at the end of the year, he had lost his voluminous notes somewhere in Spain. Or had someone stolen them? No longer knowing what to believe, he brought home but a meager part of his work. What is certain is that Pavon and Ruiz, after confiscating half of Dombey's harvest, proceeded immediately to classify and name or rename the plants.

Louis XVI and Marie Antoinette wanted to see Dombey immediately, to reward him financially and secure his future with a pension. He had returned far from empty handed: In his crates, despite the sharing of material with Spain, he had about one thousand five hundred plants, including sixty new genera. In addition he had brought back from America numerous insects, mammals, fishes, and marvelous birds, some of which, mounted on artificial trees, were sent to the queen. He had also held on to a remarkable collection of objects and vases "found in the Peruvian tombs."

Dombey was in favor and applauded by the scholarly world as a great discoverer. He was even invited to join the Académie des Sciences, but he refused. With his collections safely delivered, there was nothing to keep him in Paris. He escaped to a friend's house in the country and refused to see anyone. In 1786, still in a state of intellectual exhaustion, he moved to Lyons, where, one day, he burned all his manuscripts. His one remaining ambition was to compile a flora of Peru; it was to be the first systematic one. The flora did appear, some years later, but, signed by Ruiz and Pavon, it served to enhance the glory of Spain. The work's appearance plunged Dombey yet further into melancholy.

During the Terror in 1793 he left Lyons for Paris, to be honored there as a scholar. He was elected to the Société d'Histoire Naturelle. But that was not enough to retain him, he had to escape. He accepted with alacrity the offer of a long stay in the United States, where he was asked to study whatever he wished. He set sail from Le Havre on January 12, 1794, stopping at Pointe-à-Pitre in Guadeloupe; this was his last port of call. From Guadeloupe, where the counterrevolutionaries were winning the day, having massacred a number of people, Citizen Partarieu wrote to the abbé Gregoire on October 8, 1794:

"Citizen Representative: On reading your recent speech to the National Convention, it was clear to me that you were aware of the fate of Citizen Dombey, sent to the American continent by the Comité d'Instruction et de Salut Public. This estimable citizen is no more. His death must be attributed to the despotism of Collot, the former general of Guadeloupe and the last of the royal tigers to have escaped into the colonies from the court menagerie."

Dombey, suspected by the counterrevolutionary Collot, had tried to escape. But he was caught and thrown into jail, where he died of hunger, thirst, and fever. His herbaria were entrusted to a botanist, L'Héritier, who hid in England with the huge collection; but when he returned to France, he was assassinated.

The Revolution marked a turning point in the history of the King's Garden. Buffon had died in 1788, at the eve of the turmoil after two years of retirement in Montbard, where he completed his work on the swan, the bird that "preludes its death throes with harmonious songs." Louis-Guillaume Lemonnier, the director, had retired to the château of Montreuil, the home of Marie-Louise de Rohan-Soubise, whose estate he had turned into an exotic paradise. The Garden itself was, in 1789, integrated in a new, larger ensemble, the Muséum d'Histoire Naturelle. Fortunately, the new director, René Desfontaines, was a fine botanist. Thouin was still there, and so was Antoine-Laurent de Jussieu, a nephew of the great Jussieu brothers, now carrying on the family tradition. They were determined to stay in order to salvage the royal gardens. Thouin was elected in 1789 a surrogate deputy for the Tiers Etat: Later, as a member of the Paris Commune, he was able to convince his colleagues that the Garden, far from being a monarchical institution, was the only instrument through which a proper distribution of vegetable resources could be assured throughout France. To condemn it would amount to depriving the nation of a rich scientific heritage. He also managed to obtain and replant in the Garden cuttings, shoots, and seeds of everything growing in the royal estates at Rambouillet, Trianon, Versailles, and Marly before their sacking. He secured similar material from the gardens of the great émigrés. In 1789 a new greenhouse, planned some time earlier, was built without problems. The following year the professors and botanists formed a deliberative assembly. All in all, they saved what was essential.

Thanks to them, the last explorers and the last botanists sent off under the monarchy could still send their finds safely to Paris. The two most prominent ones were Labillardière and André Michaux.

The loss of La Pérouse was keenly felt by the scholarly world and the general public. When the Société d'Histoire Naturelle proposed to the Constituent Assembly in 1791 the formation of an expedition to search for the missing men, the plan was immediately adopted. Some navigators' accounts suggested that perhaps on some deserted, unidentified islands the men might still be alive. Although there was no proof of this, one could not hesitate to send help. Under the command of Rear Admiral Bruni d'Entrecasteaux, two vessels were soon ready to set out, the *Recherche* and the *Espérance*. A team of scholars was to travel on the *Recherche*. The group included the astronomers Bertrand and Pierson; a hydropher, Beautemps-Beaupré; an engineer, Jouvenay; a draftsman, Riché; and the botanist Labillardière; as well as Louis Ventenat, a scholar and the ship's almoner.

Jacques-Julien Houtton de Labillardière was first and foremost a Norman. He was born in Normandy in 1755 on his parents' estate. They were prominent citizens, respectful of all traditions, including that of sending their son to the college of Alençon, one of the last refuges of Jesuit influence. There he learned Latin, Greek, and philosophy. A companion persuaded him that medicine was the key to everything, particularly to adventure, and that they should both become doctors. Labillardière left for Montpellier in 1772, because that was the best place to learn medicine. A meeting with Antoine Gouan, the professor who was a friend of Commerson and Linnaeus and the inspirer of Joseph Dombey, determined Labillardière's career. He decided to devote himself to plants. When his studies were completed, he went to Paris in 1780, where he met and became friendly with René Desfontaines, who introduced him to Lemonnier, another Norman. This was to be Labillardière's Garden period, a training stage then essential for all good botanists. Lemonnier sent him to London in 1782 to study the exotic plants cultivated there. He met Joseph Banks, Cook's botanist, with whom he conversed endlessly about Australia and Oceania. Labillardière began to dream.

He returned to Paris, and in 1786 he went to Syria, a country then ravaged by war and the plague. Nonetheless, he brought back more than a thousand different plants. At length he was named botanist to the great expedition of 1791 and, like most of his companions, had few illusions about their chances of finding La Pérouse. But this was both a scientific expedition and the adventure he had been seeking. "This journey," he wrote, "merited a botanist's attention. I eagerly seized the opportunity of sailing the southern seas. Although one pays a high price to satisfy a passion for study, the varied products of a new land amply repay the inevitable hardships of these long journeys."

The *Recherche* and the *Espérance* set out, leaving a land torn by revolution, and bearing men confronting ideas. Had they never considered, or had they underestimated the difficulties they would encounter in being together for so long a period, in so confined a space as a ship, a microcosm of the French upheaval? The tension between the monarchists and the republicans, heightened by physical proximity, soon became unbearable. Labillardière was a republican, as were Ventenat, the chaplain, Riché, the draftsman, and most of the crew. Disagreements became so intense that the men almost forgot the purpose of their voyage. They reached the southern seas, but they found no trace of La Pérouse. Whenever they landed, they faced the hostile reactions of the inhabitants. In New Caledonia the natives were cannibals. Fever decimated the crew, and Bruni d'Entrecastreaux died of scurvy.

To add to their problems, there were the Dutch. Holland was then at war with republican France and so the Dutch arrested, imprisoned, and scattered what remained of the expedition when they arrived in Java. But Labillardière had already dealt with the Dutch on the island of Ambon. Like Pierre Poivre, he had faced the all-powerful Dutch East India Company. As a scholar and a republican, he deplored their methods of exploiting the natives. We may better understand the political conflicts on board the French ships if we read the following lines from Labillardière's picture of eighteenth-century colonization:

> The chief employees of the Company have the right to take from the natives, without payment, supplies necessary for their daily needs. Nothing can be more oppressive than this forced contribution, since even the hard-working native is almost certain to be left with scarcely enough to live on. Most of the inhabitants are compelled to be content to get by with easy forms of cultivation, spending in idleness time which, under another system, might have been used to achieve a certain degree of ease. The taxation authorities complete the oppression because they run the police and can impose financial penalties for their own benefit. Such penalties are fixed according to the offcers' greed and the wealth of the natives, who are often found guilty, even though they have committed no wrong.

The Dutch colonies themselves, however, were in almost as unfortunate a situation. "A stay in Batavia," noted Labillardière, "is so damaging to most Europeans, particularly during their first year, that, of 100 soldiers arriving from Europe, 90 usually die in the first twelve months. The remainder, having partially adjusted to the climate, lead a listless life." The life of a prisoner, of course, was unbearable: Louis Ventenat died in jail. Labillardière was eventually set free, but he lost everything, including his crates and herbaria. An English frigate had earlier seized the ship transporting them. When Labillardière reached France in 1795, he was empty handed. Eventually he retrieved his collections, but only after long diplomatic maneuvering. Antoine-Laurent de Jussieu had to intervene with the representative of France in London, at a time when the two nations were at war. His efforts were, however, generously seconded by Joseph Banks who appealed directly to the queen of England.

Considering the unfortunate history of the expedition, one might be tempted to believe that it had not been very fruitful. But such was not the case, at least so far as Labillardière was concerned. Despite everything, he discovered an important tree, the eucalyptus, and, for plants already known in Europe, brought back extremely complete and novel observations. He also laid the groundwork for a complete flora of New Caledonia.

He found the first specimens of eucalyptus in Tasmania. "In Tasmania," he wrote, "we were filled with admiration at the sight of the ancient forests. The finest trees in this part of the world are various species of eucalyptus. The eye was struck by the prodigious height of these trees." Most astonishing of all was undoubtedly the species called *resinifera*. Known for a long time as the Tasmanian blue gum, it produced a kind of gum and bore beautiful blue flowers. Among other species of the eucalyptus Labillardière named the *cornuta*, the *globulus*, the *cordata*, and the *virminalis*. Because the eucalyptus was known to be an important medicinal tree, Desfontaines and Thouin worked tirelessly to acclimatize it in the Garden. Some eucalyptus grow today in many sheltered areas of the Mediterranean and the Atlantic coast of France. From there it has been taken to overseas territories for transplantation. Hardy species grow as far north as Ireland and Scotland, and it has been acclimatized in the United States, mostly in California.

Labillardière published his *Relation du Voyage à la Recherche de La Pérouse*, full of observations, information, and evaluations, many not limited to the botanical domain. He studied in particular the kava (*Piper mephisticum*), and the famous, and then mysterious, breadfruit tree (*Artocarpus altilis*). Labillardière, at last, described it in detail. Of the tree's food value he wrote:

> The breadfruit tree produces for eight months of the year fruits which ripen one after the other; these supply the island-dwellers with a plentiful provision of healthy food. The fruits of the tree are almost oval in shape, three decimeters long and two wide. The entire fruit is edible, except for the extremely thin outer skin. We ate them with pleasure, giving up without regret not only our dry biscuit, but even the small portion of fresh bread normally distributed to us every day. Our cook usually served the fruit boiled in water, although it would have been far better if he had baked it in the oven.

Specimens of the tree were eventually acclimatized in the Garden.

As for the kava, it was a kind of pepper, peculiar in that it could also be served in a drink. The roots were ground to a paste, which was then infused in water. The inhabitants of Tongatapu offered some of this drink to the travelers. "But," Labillardière wrote, "it was better not to see the drink prepared if one wanted to accept the invitation of these honest folk," for the roots were ground up by chewing. This process was assigned to lower-class people who, after performing their task, spat the resulting pellet into a kind of salad bowl in which the infusion was to take place. The chewers did not themselves partake of this rich man's drink, which apparently tasted "sharp and stimulating." Labillardière was also surprised to find the hibiscus used for nutritional purposes and for making fiber rope.

His *Relation* is full of extremely valuable notes on the nature and the civilization of Oceania. But somehow the expedition had failed. The circumstances of the adventure account for many of its shortcomings. The men and their work became scattered as the journey proceeded. Another reason perhaps is that periods of great political upheaval are not favorable to scientific endeavors: Other problems occupy men's energies. The search yielded little and hope failed. Labillardière was no happier on his return. A convinced republican, he could not be a Bonapartist. Since that young general did not look kindly on the hesitant, Labillardière was seriously compromised. His position became yet more precarious when Bonaparte, eager to forward in his own way the entire field of science, revealed his own research policy in 1798 by the creation of the Institut d'Egypte. It was inaugurated by a grand ceremony at the foot of the pyramids.

Labillardière retreated into solitude. "He always missed his friends, the savages," a journalist of the day wrote. "His fierce misanthropy drove him to live on the seventh floor and to avoid visitors. During the fine season, he dwelled in a kind of hut on the outskirts of Paris." With the exception of René Desfontaines, Labillardière no longer saw anyone. He died when he was seventy-nine, in 1834.

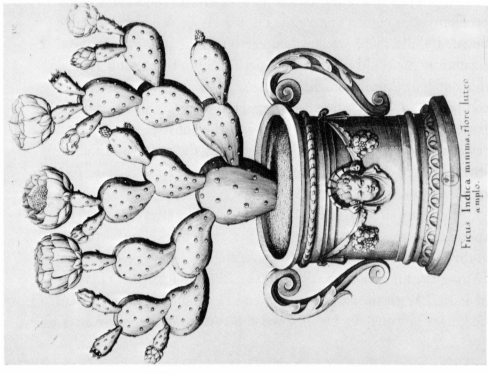

Ficus Indica minima, flore luteo amplo.

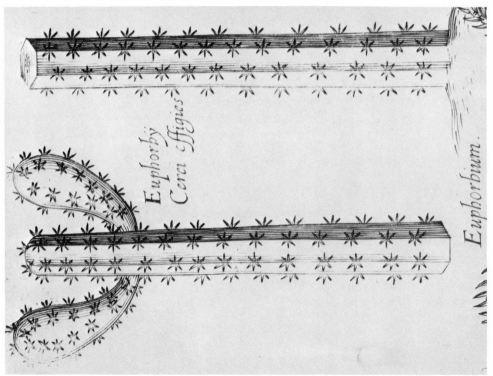

Euphorbium.

Euphorbij Cerei Effigies.

Euphorbium and Ficus Indica. Cacti became very much in fashion in the eighteenth century. (Paris, Bibliothèque Nationale. Photo by ERL)

Native village in Africa. Eighteenth-century engraving. (Paris, Bibliothèque Nationale. Photo by ERL)

[II]

Michaux *in the United States*

ANDRÉ MICHAUX was born in 1746 on the royal domain of Satory, attached to Versailles. His father, a botanist, worked for the king. He tried to solve the most difficult agricultural and horticultural problems, studying fruit trees and grains, tuberoses and other rare flowers. A most knowledgeable man, with a good social position, André's father was frequently in contact with visitors to the court. He took care to give his son a solid classical and scientific education. The young man proved gifted, even brilliant. It was easy for him to add botany to his other intellectual disciplines in order to succeed to his father's position. Rapidly he became as adept as André Thouin at working with various species.

Michaux married in 1769 when he was twenty-three, but his wife died a few years later, leaving him in despair. Fortunately for history, his neighbor was Dr. Lemonnier, physician to the king and director of the Garden. He had a great respect for Michaux, whom he had often employed in the Trianon gardens. Hoping that change might help Michaux's state of mind, Lemonnier suggested that he come to Paris to work with Bernard de Jussieu. There Michaux soon became a close collaborator of André Thouin. Both, keenly interested in acclimatization, gave their courses in the mornings and spent the rest of the day working on their "subjects." They became

126

friends. Thouin was the best acclimatizer of his day, and he knew it. He had also become the center of an extensive network of international exchanges, a project that would occupy him for the rest of his life. For him to leave Paris was out of the question.

André Michaux, however, longed for wider horizons and pointed out that they were working "too closely" together. Thouin, though unconvinced, understood Michaux's problem and arranged for him to take a trip to Persia and neighboring countries. Michaux spent three years traveling between the Caspian Sea and the Indian Ocean. He did a splendid job, but he soon realized that this voyage was not his life's great adventure. He sent Thouin liliaceae, daphnes, unknown oleanders, resedas, styrax, palm trees, and, from Shiraz, precious roses and almond trees. But he was not particularly excited by these discoveries, and so he returned to Paris, to the Garden. There, in May 1785, he witnessed the visit of Thomas Jefferson, who was hoping to induce French scientists to come to his country. Two days after his meeting with Louis XVI, Jefferson went to the King's Garden, where he asked to see "Monsieur André Thouin," who was to become a lifelong correspondent and friend. Until his death in 1824, Thouin sent every year packets of seeds to Jefferson. The future president brought with him three small bags of seeds from his country and was given others in exchange. Jefferson was enthusiastic about his visit to the Garden and the greenhouses. Thouin was deeply moved. Michaux watched thoughtfully. Americans, after this visit, soon agreed to let a French botanist prospect their forests. André Michaux was selected, and he left within a few months for the United States. He was charged with the dual mission of introducing European plants to America and sending American species back to France, a scientific import-export arrangement. France had been sending plants for years from the King's Garden to Canada, Louisiana, and the Caribbean, but now the movement of plants was to be intensified. Michaux, accompanied by his fifteen-year-old son, François-André, adapted rapidly to the pace of

127

life in the United States: "We reached New York on November 13, 1785," he wrote to Lemonnier; "on the 17th, I went to visit the New Jersey woods, where I recognized many liquidambars (*Liquidambar styraciflua*, or sweet gum), sassafras, tulip trees (*Liriodendron tulipifera*), besides various species of oak. Most of these trees are found in cool places, some of which are even flooded in winter." He added that he was sending off a crate of trees, containing, among others, 150 small liquidambars. This was soon followed by nearly five thousand seedlings and crates of seeds. Clearly, he was not wasting any time.

There were difficulties, however: "We have been unable to obtain any tree moss, necessary for packaging the young plants. We have had to make do with grass and soil. Since the soil is already frozen, our work was very difficult." The climate around New York is severe—"this country," wrote Michaux, "is almost as cold as Russia." He was disappointed, thwarted by frozen roots and trees that could not be dug up. "The problems in sending back these interesting trees," he wrote Lemonnier on another occasion, "make one all the more aware of the need for a nursery."

A nursery was indeed the only solution, both for the plants being received from Europe and for those being dispatched to France. But the New Jersey legislature did not permit the purchase of land by foreigners. An exception to the law was made, however, for the "botanist to His Majesty the King of France." Michaux became the owner of a plot of land not far from New York. Unfortunately, the climate was not ideal. Michaux was far more enthusiastic about what he had seen in a trip farther south. "One finds there," he declared, "a hundred trees for every ten in the north." The botanist to His Majesty therefore acquired a second property, near Charleston, South Carolina. The bill of sale, dated 1786, is still in the city archives. Michaux lived on this estate with his son while two other young gardeners, Paul Saulnier and Jacques Renaud, who was only 17, were left in charge of the New Jersey nursery. In the evenings, the men learned English.

Michaux was well respected because people knew that he was in the United States at the request of Jefferson and that he was the envoy of the king of France. Michaux was not just a Versailles marquis, but a strong, healthy forty-year-old who loved horseback riding. He drove men and animals relentlessly in his quest for new plants, in this rich land that he liked so much. Soon the Michaux and the hired help had prepared the hundred or so acres which they were to use half as a nursery and half as a trial ground. Now the serious work could begin. André Michaux, sometimes accompanied by his son, rode across both Carolinas, Pennsylvania, Maryland, Virginia, Kentucky, Tennessee, Georgia, and Florida. He reached the Mississippi to the west and went back up north across Canada as far Hudson Bay. One of Michaux's fondest memories was of his five-month stay with the Cherokee Indians in the southern Appalachians, with whom he traveled extensively through the forests, listening to their stories. The Cherokees always remained for him a source of information and wisdom.

This life was to last ten years, during which Michaux made of his property in Carolina a nursery soon famous throughout America. He acclimatized there everything he received from France, including camellias (*Camellia japonica*), the Indica azaleas, the gingko (*Gingko biloba*), the silk tree (*Albizzia julibrissin*), the crape myrtle (*Lagerstroemia indica*), and the Greek laurel (*Laurus nobilis*). Michaux also planted on his land all the seedlings and seeds brought back from his trips, awaiting the right moment to send plants to France. The journey across the Atlantic was long, and the plants had to be boxed or potted up at the best moment of their growth. Another consideration was the departure day of the boat. During his years in the United States, André Michaux sent back some sixty thousand plants and 90 crates of seeds. This contribution was of immense value. When he was with the Cherokees, Michaux discovered six different species of the prestigious magnolia and some new rhododendrons, notably the Flame Azalea (*Azalea calendulacea*), which he thought one of the most beautiful wild-flowers he had ever seen, and the rhododendron *Catawbiense*. He sent back to France other Piedmont natives, such as *Azalea viscosa* and *Azalea nudiflora*. Most of these were later reclassified as rhodo-dendrons by the American John Torrey and renamed: *Rhododen-dron calendulaceum, Rhododendron nudiflorum*, and *Rhododen-deron viscosum*. In a letter addressed to Thouin and Lemonnier, Michaux described entering an almost impenetrable cypress wood, declaring that "in the shade of these cypresses, where the sun's rays cannot reach, one finds in all their glory these magnolias and rhodo-dendrons. Sometimes magnolias may grow in open spaces, but the rhododendrons are at their freshest only in darkest places."

What he preferred above all, what he studied mostly, and what he became most famous for, were the great American trees. He observed the conifers, the epicea, the thuja, the white pine (*Pinus strobus*), and more than thirty other species of pine, not to mention the tulip tree (*Liriodendron tulipifera*), and the hickory. Michaux named at least three of them: *Juglans alba, J. amara, J. porcina*: Today they all are called *Carya*. He also saw twenty-six species of oak, including the red oak, maples, sassafras, lindens, and larches. Michaux made Americans aware of their country's treasures, and he helped them to draw up an inventory of their forests. He also urged them to assess the usefulness of each tree, for nature was in some danger now that the pioneers were cutting down forests in order to build settlements. George Washington had already ex-pressed concern over the problem.

Michaux was less interested in flowers, although he did send back dozens of varieties of asters, bignonias from Virginia, agaves, and an orange lily which he named *Lilium carolinianum* but which today bears his name, *Lilium michauxii*. He worked mainly from

Europe to America, bringing over oriental flowers acclimatized in Paris. His enthusiasm was infectious. Botanical societies were formed all over the Unites States, and "well-educated" young men, as he called them, formed a group around him. They laid the foundations for the American school of botany, which has been most active ever since.

Michaux also set a fashion in France. The French taste for plants had not extended to trees, with the possible exception of fruit trees. Trees were simply part of the decor, in parks or along the roads. As for forests, those green masses had never had a good reputation. They concealed too many ancient terrors. But André Michaux managed to communicate his enthusiasm to such an extent that at the end of the century arboreta were started with American species that he sent back. A few collections, public and private, became veritable tree museums. (Some of these still exist; they are listed in the appendix.)

This fondness for trees was an idea suited to the age, appropriate to such civilized men as eighteenth-century Europeans believed themselves to be. Louis XVI was particularly fond of hickories and the fast-growing American beeches (*Fagus*), but with the general public pines were most popular, second only to magnolias. Pines were beautiful, healthful trees, and they came from distant lands. If positioned carefully, they adapted well to the French climate. The white pines (*Pinus strobus*) were particularly sought after, as was the Douglas fir (*Pseudotsuga taxifolia* or *douglasii*), both as an ornamental tree and for cabinetmaking.

It is impossible to list all the names of the other trees that became popular; the American Larch (*Larix*), for example, stronger and larger than any known before; the hemlock, a handsome conifer; the swamp, or bald, cypress (*Taxodium distichum*), most striking when planted beside a river, and the marvelous sequoia, said to be the tallest of all trees. The size of the American thujas amazed everyone. Indeed, the chief characteristic of the

American trees seemed to be their size. The American ashes (*Fraxinus*), for example, with their lovely scented flowers, saved from certain degeneration the frail Caucasian ashes of Europe. The same improvement took place in linden trees (*Tilia*), the various species of the New World helping to strengthen those of the Old. We know that twenty-seven species of maple were brought back by Michaux, as well as balsam poplars (*Populus trichocarpa*), the sassafras, and endless magnolias. The Cote d'Azur is still resplendent with the strange yuccas acclimatized from America, whose spikes of greenish white flowers are pollinated by a female butterfly.

André Michaux returned to France in 1796. During his absence the French Revolution had taken place. When he returned to Paris, he no longer recognized his world.

Michaux, however, was the hero of the day, feted by all, even if he no longer felt at home in the Garden. He set to work with Thouin, eager to watch over the successful adaptation of the plants from America and their distribution throughout France. Government policy was then most favorable to reforestation. François-André, his son, continued to supervise the link with the United States, where Paul Saulnier had remained in charge of the New Jersey gardens.

At this point another figure came on the scene, the new director of the Garden, René Desfontaines, who held his post as director for forty-five years. He was born in Brittany in 1750, the son of a well-off family of landowning peasant farmers. Desfontaines studied in Rennes, where he was successful in all disciplines. Finally he turned to medicine. Medical study was at that time complemented by a period spent in the lecture halls of the King's Garden. There, Buffon, Daubenton (who taught zoology), and Antoine-Laurent de Jussieu dazzled the young Desfontaines and determined the direction of his future career. He too was to become a botanist.

Desfontaines had spent a great deal of time in the Garden as a student. He was familiar with its layout, contents, archives, and organization. With his natural gift for organization, he could not but notice certain gaps, even oversights, particularly in the classification of the herbaria that had been sent or given to the Garden by all the botanist-travelers. He began to dream of a new classification of the volumes which would make possible a complete listing of everything that had not been planted in the Garden but had appeared only in the herbaria. Such a compendium would certainly contain some surprises and would encourage new study and analysis. It would also be a fine way to pay homage to those who had assembled the herbaria, often at the risk of their lives. Not all the travelers had been able to bring back live plants, but all had incorporated their knowledge and experience into these volumes. Desfontaines began work on the task.

He was not a man dedicated exclusively to raking out dusty engravings and sorting dried plants. He had been made a member of the Académie des Sciences in 1783 because of his study on the *Irritabilité des plantes*, and he had also traveled to Barbary, as Algeria and Tunisia were then called. Yet the Garden paths were of sufficient scope for Desfontaines; the glass houses were his true home. After much labor he completed his project, and he set up what is still called the National Herbarium of the Museum. In 1794 he opened the Museum library, now enriched with the material he had collected, to the public. Every day, he worked on the plants themselves with Antoine-Laurent de Jussieu and Thouin. He had been elected director of the Garden by his colleagues, who included now, in addition to those already mentioned, the naturalists Jean-Baptiste de Lamarck and Etienne de Lacépède, and the zoologist Geoffroy Saint-Hilaire.

This new regime suited Michaux less well than the New World. Much distressed him, in particular the blind destruction of the vast royal plantations. It was difficult for a man who had freely roamed the wide open spaces of America to follow directives from others. He was also troubled by severe money problems. The monarchy owed him five years' salary, but the revolutionary government refused to honor the king's debts. Nor was there any chance of his obtaining a fresh subvention. He was appointed to an agricultural commission, but the post bored him. There were causes for satisfaction, however. In every forest of France there were trees that he had brought back. The fact was known and acknowledged, but Michaux, in full possession of his physical faculties, and now in his early fifties, still longed to travel.

It might seem that his desire for change was chiefly responsible for his accepting a position as botanist to Captain Baudin's expedition to the South Seas. But he had another idea at the back of his mind: He wanted to go to Madagascar. He said nothing of his intention. He had decided to leave the expedition in Mauritius and to go on from there to the "botanists' paradise," where he hoped to continue Pierre Poivre's work. His dream was to set up on the "Grande Îsle" the equivalent of Poivre's Grapefruit Garden in Mauritius to supply France with the new garden's products, and he wanted above all to compile the flora of Madagascar. Many had dreamed of completing such a task, but none of them, from Flacourt to Poivre and Commerson, had ever managed to finish it.

Michaux left with the Baudin expedition. As he had planned, he left the team in March 1801 when they reached Mauritius and after six months there proceeded to Madagascar. He landed in Tamatave, but he settled in an unexplored region away from the coast. In Isantrana he found an isolated plot of land on which, as he had in America, he built a cabin. There, working alone and without respite, he laid out and began to plant his garden. What happened next is unclear, but probably he caught a fever. Colleagues who had followed him thought he was off on an expedition and did not immediately go to investigate. In fact, he died in his wooden house, alone as usual during this latter part of his life, in November 1802. Around him grew mango and guava trees, avocados and lichees, medlars, and tea and coffee bushes. Not a single manuscript or note was found. Had he left none, or had they been taken? Again, the botanists' paradise did not reveal its secrets. Michaux's book on American oaks, the *Histoire des Chênes de l'Amerique*, beautifully illustrated by Pierre Joseph Redouté, had been published in 1801. His comprehensive work on the flora of North America, the first ever written, *Flora Boreali-Americana*, was to be completed by his son and published after his death, in 1803. It was also illustrated by Redouté.

Baudin's expedition, with which Michaux had made his last voyage, was itself doomed. Indeed, with Baudin as its leader it never had a chance to succeed.

Baudin was born in 1750 on the Île de Ré, where the sea was part of his life. In 1786 he was second lieutenant on a sailing vessel. Shortly after, in India, he captained a ship flying the Austrian flag. His task was to conduct research into natural history on behalf of the Hapsburg emperor, Francis II. He made a long voyage to the West Indies. He returned to present to France the collection he had accumulated for Austria. This strange exploit was rewarded by a captain's rank. The cargo was so rich that the Muséum had to build a special glass house, the Baudin glass house, for the plants he brought back. He made a favorable impression on the Muséum scientists and also upon Bonaparte, who wanted to find out about the Antipodes and to learn whether, as was sometimes claimed, Australia was divided in half by an arm of the sea. He also wanted maps of the region. Accordingly, two corvettes, the *Géographe* and the *Naturaliste*, were armed and entrusted to Baudin. The boats were to carry a large group of scientists, twenty-three in all, from every discipline. Among these were: in zoology, René Maugé, Désiré Dumont, and François Péron; in mineralogy, Charles Bailly and Louis Depuch; in hydrography, Charles-Pierre Boulanger; in astronomy, Frédéric Bissy and Pierre-François Bernier; in painting and draftsmanship, Charles Alexandre Lesueur, Nicolas Martin Petit, Jacques Milbert, Louis Lebrun, and Michel Garnier; in gardening, Antoine Sautier and Antoine Guichenault; and, last, in botany, Leschenault de la Tour, Bory de Saint-Vincent, Anselme Riedlé, Jacques Delisle, and as we have seen, André Michaux.

The group left Le Havre on October 19, 1800, the last autumn of the century. The journey began in a festive atmosphere: "We swear unity between officers and scholars. . . . Those who break this vow will be punished by the others' scorn." The words had a prophetic ring. The two ships were well equipped; part of the holds had been specially built to preserve living plants brought from France and to raise seedlings sprouted from seed gathered in the course of the trip. Until the end, Riedlé watched over this floating garden, leaving in his papers an account of everything sown.

The expedition reached the Canary Islands without any difficulty other than an increasing disharmony on board. The captain showed scant sympathy for the young scholars, whether from misplaced pride or from irritation at his intellectual passengers. He was so authoritarian that those on board held long discussions as to his fitness for leading such an expedition. Baudin observed his orders strictly. He never tampered with the sailing schedule, even when it did not coincide with the needs of scientific observation. Intent on their research, the numerous young scientists had no desire, of course, to be ordered about by the captain. Eventually, when Péron and Bory de Saint-Vincent began voicing criticisms of Baudin, some officers and part of the crew joined them. This exacerbated the situation. After landing in Tenerife, they did little work, because they were too busy arguing about trivia. Baudin wanted to regulate and control the scientists' excursions. The only person allowed to land when he wished was Riedlé, who made a plentiful collection which he arranged for a Spanish vessel returning to Cadiz to forward to the Muséum.

When they set off again, Baudin committed his first navigational error. Instead of moving out into the open sea, he hugged the coast of Africa. This decision resulted in four months of exhausting and infuriating sailing. At times they were motionless in the equatorial calms; at others, they were battered by heavy storms. Tensions mounted. The passengers did a lot of writing, most of it highly uncomplimentary to the captain. Riedlé, recovering slowly from an accident in Tenerife, watched conscientiously over the seedlings in the hold. Plum, apricot, apple, and pear trees bloomed, olives grew fast, and lettuce thrived.

During this part of the voyage scurvy plagued the travelers: Bory de Saint-Vincent felt the first symptoms. In March 1801, at Île de France (Mauritius), Michaux left the expedition as he intended, and the first desertions took place. Despite his illness, Bory de Saint-Vincent continued to explore. On Île des Tonneliers, he discovered the boraginaceous shrub *Tournefortia* and some pink periwinkles (*Vinca rosea*). At Camp Masque he gathered orchids and studied the raffia palm, or Madagascar raffia (*Raphia ruffia*). Baudin, meantime, did not receive the third vessel promised to him by Paris and supposed to join them at this point. The provisions available were insufficient. Just as they were leaving, some scientists —Bory de Saint-Vincent, and Milbert, by now convinced that the expedition would never come to anything while headed by Baudin, decided to follow Michaux. After this first major setback forty more men, including many crew members and some officers, deserted the expedition.

The remaining crew set sail and arrived within sight of Australia on May 27, 1801. A landing was impossible, except on a few small islands. Since they had missed the best season for botanical work because of earlier delays, it became impossible to fulfill the program proposed by the institute. Despite continuing discontent, they explored, near Cape Leeuwin, a relatively barren area on which Riedlé found much to do. "Our gardener," wrote Péron, "gathered a rich harvest of new plants, which he paid for with seedlings of corn, maize, barley, and fruit trees. These touching exchanges should always be used as the basis for friendships between peoples."

What happened next was described later by Jules Verne in his *Histoire générale des grands voyages et des grands voyageurs*: "Because the violence of the sea had driven the two vessels away from the coast, two crew members were forced to spend several days on land, with nothing to drink but brackish water, able to find neither animals nor birds to kill. Their only food was a kind of saxifrage. This plant provides a large quantity of soda, and contains an extremely bitter juice. They were obliged to abandon a rowboat beached by the waves, as well as guns, sabers, cartridges, cables, tackle, and many other objects."

They weighed anchor again on June 10, but then immediately lost sight of the second boat, the *Naturaliste*, commanded by Hamelin. To everyone's surprise Baudin decided to continue alone. The two vessels did not meet again until two months later, in Timor. On July 20 the *Géographe* was in the Bay of Dogs, where Baudin conducted some cartographical surveys, but he refused to allow anyone to land. "I was mortified," noted Riedlé, "not to be able to study the vegetation and gather new species of plants." The botanists grew increasingly annoyed at the captain's capriciousness. The other scientists, geographers, astronomers, and hydrographers could work on board ship. A short time later, on another island, they almost lost Péron. Baudin, indifferent to the scientists' plight, told the officer in charge of the search party: "Go ahead, but be back by nightfall, with or without your rock collector."

The scientists took a much needed rest in Timor from August to November. Several men had to be hospitalized for scurvy. When the *Naturaliste* finally rejoined them, they were able to relax, but not for long. Riedlé died of exhaustion a few days after their arrival. Then, on November 6, Antoine Sautier, the gardener, died of tetanus. "Quite apart from our grief at his death because of his qualities as an individual," wrote Baudin, "I soon realized to my sorrow that the botanical part of our research would not be completed, for I had unfortunately lost the two men who were working so zealously in that area."

Baudin himself had been sick since September, complaining of numbness in his legs and a high fever. In November 1801 they set sail for Van Diemen's Land, the modern Tasmania, and arrived there in January 1802. Illness claimed another three lives, including that of Maugé, the zoologist. In May the situation became unbearable; food was short and water was rationed. Only four men were still fit to navigate the ship. Baudin's condition had worsened, making it increasingly necessary for his lieutenant, Henri Desaulses de Freycinet, to replace him. Having lost the *Naturaliste* again, Baudin finally resigned himself to putting in at Port Jackson. They arrived at the end of June and remained there for five months.

The expedition had lost so much time that the two surviving botanists, Leschenault and Delisle, and the gardener, Guichenault, had to work fast. "Governor King," they wrote in their report, "had planted in front of his Italian-style house a sort of Grapefruit Garden, in which we admired some splendid Norfolk Island pines (*Araucaria excelsa*), some columbias, Asian bamboos, Portuguese orange trees, and Canary figs." They were back in a more familiar environment at last.

When the *Naturaliste* eventually caught up again with them, Baudin decided to send the vessel back to France with the scientists' collections. Péron and Lesueur were directed to prepare the shipment for the Muséum—33 giant crates containing 40,000 plants, 200 birds, 68 stuffed animals, and 1,500 drawings. To carry provisions and replace the *Naturaliste*, Baudin bought a schooner, the *Casuarina*. They bade farewell to the homebound ship and continued on, having transferred all the healthy men to the *Géographe*. On November 18, 1802, they were once more in difficulty among unknown islands and hostile natives. On February 8, 1803, Péron began to draft a description of the expedition, which would pay tribute to those who had died. But the journey was by no means over. They sailed past island after island. They continued their work without Leschenault who, too sick to travel, remained in Timor, where Baudin had stopped a second time. The expedition then made a new attempt to land in Australia. In the course of sailing round the continent, they had at least ascertained that it was not divided in two, a fact of interest to Bonaparte. Pierre-François Bernier, the astronomer, died on June 6. Baudin, still eager to reach New Guinea, refused to retreat. Contrary winds, however, forced him to take the route back to Île de France. "One can imagine," wrote Baudin, "the impression made by this change of direction, which no one expected, because no one ever knew where I was going, what I intended to do, or what I actually was doing during the journey."

Île de France was Baudin's last port of call. He died there five weeks after he arrived. If they had not been constrained by a sense of propriety, the travelers might have rejoiced at his death. But they began to talk more openly: "Baudin's authoritarianism," Guillaumin, the historian, was to write, "his pride, violent temper, scorn for scientific work, and indifference to the sufferings and lives of others caused his death to be seen as a manifestation of immanent justice." On March 24, 1804, the *Géographe*, commanded by Freycinet, entered the port of Lorient, in Brittany, bringing back 70 crates containing 200 species of useful plants and seeds of 600 species. In 1807 Freycinet and Péron published the first volume of their *Voyage de découvertes aux terres autrales*. But scholars in Paris were horrified by the death toll. In addition to Riedlé, Sautier, and Maugé, the mineralogist Louis Depuch had died on Île de France, on June 6, 1803. The astronomer François Bernier had died at sea on June 6, 1803. And André Michaux, who had also sailed with Baudin, had died in Madagascar.

[12]

Raffenau-Delile *in Egypt*

Bonaparte's imagination was always haunted by the memory of Ancient Rome. He was obsessed with certain words —*republic, code, consul, triumphal arch*—and he had read the classics from Livy to Tacitus. But after his conquest of Italy, the time was not ripe to bring Rome to Paris, to become a new Caesar. The young general turned his gaze to Egypt, its vast desert lands and its gigantic pyramids. He also wanted to close the English route to the Orient and to oust the British fleet from the Mediterranean.

To take possession of Egypt seemed to Bonaparte logical. He assembled 33 warships, 232 transport ships, 2,000 cannon, 33,000 soldiers, 175 engineers, and, for the progress of knowledge—because the Republic should be universal—an Areopagus of 200 scholars. Through conquest Bonaparte would become the first Egyptologist of modern history. The chemist Claude Berthollet, the mathematician Gaspard Monge, and the naturalist Etienne Geoffroy Saint-Hilaire, all charged with recruiting, screened the newly founded Ecole Polytechnique, Ecole Normale, and Ecole des Mines. Their task was difficult: They had to find young men interested in the project and ready to view themselves as soldiers. Recruits had to show complete esprit de corps and, initially, be prepared to classify their activities among countless military secrets. The scholars eventually found were for the most part about the same age as their general: under thirty years old. Like him, they had nothing to lose but their lives. Because an expedition of this kind without a botanist was inconceivable, Alire Raffeneau-Delile was appointed to the position.

Raffeneau-Delile was a child of the Revolution. He was ten years old when it began. He was born on January 23, 1778, in Versailles, where his father held the position of equerry to the king and his mother was in the service of the queen. The boy received an education appropriate to his parents' position. But the Revolution entered his life on July 14, 1790, on the Champ de Mars, where he admired The Fête de la Fédération. That day, temporarily, he felt that he was no longer "stifled," a feeling that always troubled him. He refused to go back to the smothering atmosphere of his secondary school, a place as yet untouched by the new ideas in the air. His father did not object, but he kept him in Versailles and hired the abbé Cotereau as his tutor. The father was hoping to shield the boy from the upheavals in Paris. But the real danger for his son, the longing for freedom and the material for dreams, lay there, before his eyes, in the gardens and glass houses of the Trianon. The boy was always there, roaming around the plants, leaning over the rarest flowers. Dr. Lemonnier and Dr. Brunyer, physician in the Hospices of Versailles, befriended the inquiring youngster. Both men became interested in him. Brunyer took him on as a nonresident student at his hospital. Raffeneau-Delile began to divide his time between medicine, for which he was obviously gifted, and conversations with Lemonnier, who revealed the plant world to him. He met important people, Thouin, Jussieu, and Desfontaines. The last-named finally convinced him to choose botany. Desfontaines had only traveled once, to Barbary, but his travel stories delighted the young man, who set to work to become a botanist. When the Egypt Expedition was assembled, Raffeneau-Delile joined it as botanist: He had just turned twenty. He was to work with the draftsman and printer Pierre-Joseph Redouté, the illustrator of Michaux's books on American flora.

The expeditionary group assembled in Marseilles and then boarded ship in Toulon. Under the command of Brueys d'Aigallier the fleet sailed off on May 19, 1798. The captain was worried. He was aware that Nelson, the dreaded Englishman, was roaming the Mediterranean. Bonaparte, however, was unperturbed. He was certain he could cross the Mediterranean, as indeed he did. So great was his confidence and so eager was he to demonstrate it, that he spent the trip conversing with his scholars. These rambling discussions covered Plutarch, Homer, the Koran, and the stars. En route, the expedition took possession of Malta. They reached Alexandria safely and entered Cairo on July 23.

Bonaparte's scientific and artistic venture began officially on August 20, 1798, the date of the inauguration of the Institut d'Egypte, which held its first session at the foot of the pyramids. The president of the institute was Gaspard Monge, the vice president, Napoleon Bonaparte. Its task was to "research, study, and publish the natural, industrial, and historical facts about Egypt." In case of need, they had brought with them a sizable library and a small printing press.

In Cairo the scientific mission was housed in two palaces and some villas. They set up a menagerie in the garden of one villa. In the grounds of another Raffeneau-Delile at once began to create a botanical garden. The scholars' enthusiasm was at its height as they traveled the conquered territory. The artillery captain Boussard discovered the famous Rosetta stone, which would lead Champollion to decipher Egyptian hieroglyphs in 1824. Raffeneau-Delile began to turn his attention to the Nile lotus.

The period of lightning victories soon passed. Egypt was conquered, but Nelson destroyed the French fleet at Abukir. Their way home blocked, the scientists found themselves stranded in Egypt.

But they were busy with their work and did not care. Raffeneau-Delile in his garden watched over the plants he had brought back from his excursions, and he waited for the day when he could load them on board ship. He searched for more plants, riding a dromedary, which was for him a nightmarish experience. He could not get used to sitting on a hump. His aversion became one of the standing jokes of the mission.

What interested him most was the lotus, sacred along the banks of the Nile. The lotus was carved on the capitals of the columns of the great temple of Thebes, the modern Karnak. This flower, of the Nymphaeaceae family, was said in legends to be the bride of the Nile, because when the river rises, it alone is still visible above the water, as if it were celebrating the fertilizing flood. This miracle was likened by ancient Egyptians to that of the creation of the world. According to an ancient myth, the lotus one day emerged from the primeval ocean, bloomed, and when it opened, it disclosed a child carried in the calyx. The child was Ra, the sun-god, father of the gods. A different version claimed that the flower revealed a scarab, the symbol of the sun. The scarab metamorphosed into a male child, who began to cry and his tears gave birth to the human race. Other tales evoked the fabled land of the Lotophagi, or Lotus-eaters, a kind of Eden situated somewhere "between Arabia and Barbary," whose inhabitants fed solely on the plant said to induce an excessive indolence. Ulysses' companions forgot their native land when they tasted the lotus.

Such a species had to be approached with respect. Raffeneau-Delile was filled with awe at the mysteries of a plant whose seeds could lie dormant, as if dead, for over a century, then suddenly send up dazzling flowers. Unlike its cousins from Asia (*Nelumbo nucifera*), America (*Nelumbo lutea*), or all the already known nymphaeas, the sacred lotus (*Nymphaea lotus*), with its pink flowers and broad leaves, was always difficult to grow outside its native habitat. It had been acclimatized in a few pools, but it lost much of its strange original beauty. No one had ever succeeded in transporting the peculiar alluvions of the Nile waters on which it feeds.

Raffeneau-Delile observed that the water lilies' flowers opened and closed each day at the same precise instant, as if they obeyed some mysterious inner clock, and found that the floating flowers formed a family, a separate vegetable world, which could provide valuable information about the power of water itself. He was fascinated by the brilliance of their flowers, the silent strength of their leaves, which suddenly rose up out of the glaucous water in tight, glossy carpets to offer brief, frightening glimpses of the murky aquatic germination processes. Soon he became interested in all the aquatic flora of Egypt and discovered many species. But he had to drag himself away from this magic, and climb back on the dromedary hump and ride off across the sand to study desert succulents, date palms, and almond trees. He observed the papyrus, another sacred plant, which was said to be used for making boats for the gods. Maupassant has called it the "secret guardian of thought," because from its stems, cut into strips and glued together, the ancient Egyptians made the first books.

All this, and much more, was described with loving care by Raffeneau-Delile, while Pierre-Joseph Redouté was busy drawing. Unceasingly, he gathered, replanted, named, and commented, hoping to return with a complete herbarium. He studied everything. Raffeneau-Delile examined Egyptian corn with care, particularly the Pharaohs' bare-grained corn, with its great yield and tender kernels. He discovered a local rye grass, the darnel (*Lolium temulentum*), which seemed to him to have a disastrous effect on surrounding crops, and he took notes on the coronilla, a leguminous plant from which a laxative was obtained. He lingered over flowers, seeking the origin of some unknown roses. He gathered blue everlasting flowers, asphodels (*Asphodelus acaulis*), giant nasturtiums (*Tropaeolum*), and a local dianthus with peppery fragrance, which he was eager to introduce into France, because its beauty surpassed that of previously known species. Raffeneau-Delile also found violets (*viola*) with a different perfume and mignonettes (*Reseda odorata*) so sweet-smelling that the Egyptians called them "love grass." Euphorbias occupied an important place in his transplanting: He watched over five new species. He was convinced that they would make fine apartment and greenhouse plants.

Raffeneau-Delile concentrated entirely on his plants, with never a thought of the war, which had taken a dramatic turn. The defeat of Abukir had sounded the death knell of Bonaparte's Egyptian dream. Had the general spent too much time with his scholars and not enough with his soldiers, one of his officers asked reproachfully? Whatever the reason, the English were the masters of the Mediterranean. The Turks now entered the conflict and forced the French to fight in Syria. In Jaffa the army caught the plague, and Gen. Jacques Menou, despite his important responsibilities, converted to Islam. Bonaparte vanished. He returned secretly to France and left the command of the army to General Kléber. He took Monge and Berthollet with him, but the others remained in Egypt. The situation worsened rapidly. Kléber was assassinated during an anti-French riot in Cairo. The city was besieged by the English, and the French could no longer count on Abdallah Menou, as the general now styled himself. He was busy on a love idyll with his young Egyptian wife. The scholars had to fight. Geoffroy Saint-Hilaire and the doctors Larrey and Desgenettes, Labillardière's old friend, showed the greatest composure. They struggled to save everything possible from the accumulated scientific material. In the end the English captured the city and the Rosetta stone.

Raffeneau-Delile fought well, but he had to leave Cairo, abandoning his cuttings and herbaria. Later he assembled a small caravan with a Muslim friend and came back to the city which he entered at night. He loaded up as much as possible, though he had to leave a great many things behind. Eventually, these recaptured cuttings and herbaria were placed on board the brig *L'Oiseau* which, in accordance with the terms of the evacuation agreement signed with the English, was to repatriate a final group of French scholars. But they had scarcely set sail before they were intercepted by English ships and searched from stem to stern. All Raffeneau-Delile's material was confiscated. To justify their actions, the English invoked Article XVI of the Alexandria capitulation, a clause that permitted them to classify Egyptian flora as art objects. A great row followed while the ship was held in Alexandria. Raffeneau-Delile went from one vessel to another, trying to convince the English commanders to let him leave with his plants. He was unsuccessful: The plants were Egyptian, and therefore now English. Finally, the English admiral who had defeated Jacques Abdallah Menou was told the story. Raffeneau-Delile was brought before him. He protested that he wanted his plants, would not leave them for anything in the world, would rather become a prisoner in order to remain with them, and was prepared, if need be, to go with them to London.

The British admiral could not but appreciate these feelings expressed with great aplomb by a young man of twenty-two. He gave his orders, and *L'Oiseau* finally left Alexandria for Marseilles. There Raffeneau-Delile immediately organized the transfer of his treasures to the King's Garden. The crates were triumphantly received by the delighted Thouin, Jussieu, and Desfontaines. Raffeneau-Delile, however, did not share their elation. The young hero seemed weakened, melancholy, and despairing. He was haunted by the knowledge that Abdallah Menou had armed the scholars and that some of them had died in battles doomed in advance. Was that, he wondered, what was meant by scientific adventure?

One of the first pieces of news awaiting Raffeneau-Delile on his arrival in the capital was the coup d'état of the eighteenth Brumaire. Rome was in Paris. Having chatted daily with a friendly young general in love with science, progress, and reform, Raffeneau-Delile returned to find a First Consul bent on reigning. In vain the consul received the scientist, heaped praise upon him, and urged him to publish his findings. The young man resisted. He was worried, nervous, and unstable. He worked, but without enthusiasm. He had no choice, however, for his friends at the Institut d'Egypte, less unhappy than he at history's strange reversals, were extremely active. The Garden was seething with excitement, thanks to the efforts of Lamarck and Chaptal, who were teaching natural science there. Raffeneau-Delile could not remain on the fringes of such a movement.

Finally he published a remarkable study of the lotus and drew up plans for what was to be his great Egyptian flora. He also published a fine album on the liliaceous family with illustrations by Redouté as well as a study on the doum palm (*Hyphaene thebaica*) of Upper Egypt and a memoir of the plants cultivated in that region. But nothing helped him to find inner peace. He dreamed of his African gardens where, as he put it, "lemon and orange trees, date palms and sycamore figs, grow at random. The mixture of these trees, their leafy vault, impenetrable to the sun's rays, and the flowers scattered here and there in these groves make a delightful shade. When the air burns and man pants for coolness, with what pleasure does he breathe the cool air of these bowers, at the edge of the stream watering them." He still dreamed of the floating mysteries of the glaucous aquatic universe.

In its place he had a library, a laboratory, and an office, with herbaria for souvenirs. His youth was fading, his spirit broken. He continued to work, however, agreeing to write a study of the banana trees (*Musa ensete*) he had seen in Rosetta and Cairo which were now being experimentally grown in the Garden hothouses. He began this task because it was within his field, just as he helped Thouin experiment with Egyptian cotton and rose grafting. But nothing dispelled his melancholy. Jomard, the expedition's geographer, tried to persuade Raffeneau-Delile to work with him, and Geoffroy Saint-Hilaire paid him frequent visits, but these were to no avail.

In 1803 he received from Bonaparte, now First Consul for life, a strange proposition. This was a year of important innovation in Paris. Symbolically, the steamship built by the American Robert Fulton was being tested on the Seine. After creating the Ecole Militaire of Fontainebleau, the First Consul reformed the main mechanisms of the state, and he set up an entirely new administration. Bonaparte needed new men; he may also have been following a suggestion from one of the members of the Institut d'Egypte when he turned to Raffeneau-Delile, or perhaps he sought only to bring the scientist out of his neurotic state by interrupting his botanical career for a time. He named Raffeneau-Delile subcommissioner for commerce in North Carolina.

Raffeneau-Delile was delighted to go to the United States, where, ever since André Michaux, French botanists had been held in high esteem. Thomas Jefferson, now third president of the United States, invited him to dine. Raffeneau-Delile met again Jérôme Bonaparte and the French ambassador, General Bernadotte. He settled in Wilmington and tried sincerely to become interested in trade. The countryside was charming, its handsome trees reflecting in the water "the image of their leafy arches." Raffeneau-Delile began at once to plan the foundations for a flora of North Carolina, a work he was to leave unfinished. But despite a violent hurricane, followed by a fire which destroyed half the city, he was bored. Finally he made a strange decision. He left Wilmington and went to Philadelphia, where he returned to the study of medicine, taking courses at Pennsylvania Hospital. He moved then to New York City, where, having received his degree in 1807, he became a specialist in pulmonary phthisis. Now that he had regained his self-confidence, he returned to France.

There the situation had changed yet again. The consul for life was now the emperor. Raffeneau-Delile was no longer concerned. He had returned to pick up again his work as a botanist, and he set about doing so. First, however, he retook his medical exams because his American degrees were not acceptable in France, and he opened an office. He married and had his first child, a son. Returning to the Muséum and his herbaria, he began to publish his works in 1808. Ten calm, productive years followed, during which some superb roses were born of his Egyptian roses, themselves originating in Persia.

The year 1819 marked a turning point. Raffeneau-Delile accepted the offer of the chair of botany and the directorship of the gardens at Montpellier. He occupied this position for thirty years. Finally he was at peace with himself. The tradition of the gardens begun by Guillaume Rondelet had continued down the centuries. This garden of medicinal plants was the richest in France, and the stay-at-home scholars who kept the school going had expanded it considerably. The department of herbaria was outstanding. Raffeneau-Delile's predecessor was Pyrame de Candolle, a Swiss scholar who had come to Montpellier at the request of the Académie des Sciences, and he had taught there for eight years. Candolle had given the school a new dimension, enlarged the garden, created a forestry school, planted numerous trees which were classified by families, and devoted part of the land at his disposal to an impressive collection of vines from all the Mediterranean countries. He had also constructed a huge glass house that contained bamboos, nettle trees (*Celtis australis*), turpentine trees (*Pistacia terebinthus*), lentiscs (*Pistacia lentiscus*), cistuses, junipers (*Juniperus*), and orchids. What Raffeneau-Delile most admired in his predecessor's work was the system of drains and watering by pipes which he had perfected. Inspired, the new director created pools in which to raise and study his Nile lotuses, his blue nymphaeas, water irises, and unpredictable water hyacinths. His first pool, the *bassin des Nelumbos*, is still there, surrounded by the palm trees he planted.

During the thirty years he spent in Montpellier, Raffeneau-Delile produced essays, studies, and observations. He undertook complicated experiments on the most difficult plants. Surrounded by understanding admirers he might have seemed content. But one day in 1850 his friend Dr. Lallemand was asked to go to Egypt to care for the khedive. Raffeneau-Delile decided to accompany Lallemand. He wanted to see Egypt, the Nile, and the gardens again. When he reached Marseilles, however, he realized that he was too ill to continue: From the docks, he watched the boat depart. Exhausted by this last effort, he returned to Montpellier and died there on July 5, 1850. In the *bassin des Nelumbos*, the sacred lotuses preserve the memory of Raffeneau-Delile's pyramids and mirages.

Watercolor of a sassafras leaf. From André Michaux's Histoire des Chênes
. . . de l'Amérique Septentrionale *(1801).*
(Charlottesville, University of Virginia Library)

Bonaparte visits the French countryside. (Photo by J.-L. Charmet)

[13]

Bonpland *and* Humboldt

FOR almost a quarter century the Revolution and the empire borrowed their images and symbols from Greco-Roman antiquity and the Orient. The fashion lasted for awhile, faded, and was finally rejected. Chateaubriand first, during the Revolution, turned taste to the West. Everything about the Americas was new or little known. Scant information was available about Central and South America because the Spaniards had set up an impenetrable barrier around those areas. The little that was known, mainly the stories told by Joseph de Jussieu, La Condamine, and Dombey, who had been there, lent credibility to the legend of the Green Hell, the accursed continent. However, the reports of rubber, quinquina, *maté*, cola, and, above all, of orchids, suggested that it was also a land of fabulous riches. In the Garden and elsewhere scientists dreamed of penetrating this forbidden vegetable kingdom. They even welcomed suggestions of danger, tired as they were of the "Age of Reason." The academies and the government, less moved by such dreams, had no plans to send missions across the Atlantic. Would-be explorers were forced to bide their time and hope for a miracle.

In 1798 Friedrich Heinrich Alexander von Humboldt arrived in Paris. He was twenty-nine and he had received the best education Germany could provide for its students, one that offered the broadest range of knowledge then available. But Humboldt chose Paris, because, as he said, he felt constricted by his family and country and he admired the French philosophers. The son of a noble Prussian family, Humboldt was born in Berlin on September 14, 1769. His father, Major von Humboldt, occupied an important position with the king of Prussia, a post that left him little leisure. He provided tutors for his children, Alexander and Wilhelm. Because he was passionately interested in the natural sciences, he made sure that the tutors were the best teachers and scholars of the day. In addition, Goethe, a family friend, exercised a profound influence on young Alexander. The boy learned mathematics, philosophy, archaeology, astronomy, mineralogy, vulcanology, geology, and international commercial law as well as English, French, and Spanish. Hoping to make his son a government engineer, the father sent him to the mining academy. But the minerals yielded by mines and the extraction processes interested Humboldt far less than the fossils which he discovered in the ore. These provoked long reflections and brought him back to his true passion, the mysteries of natural history. He resigned his post and went to Paris to meet those who, in and around the Garden, had devoted their lives to the study of nature. In Paris he soon attended the aged Bougainville's salon, where the regular guests discovered that Humboldt, by now the chief attraction at these gatherings, differed markedly in personality from the average scholar. Scientists in France had already become specialists, brilliant but limited to their particular discipline. Humboldt, on the other hand, had that synthesizing turn of mind that, most notably with Hegel, was to be the major contribution of German philosophy. "The physical sciences," he declared, "are bound together with the same ties as those linking natural phenomena." Remarks such as these soon made Humboldt a man to be listened to, particularly when he spoke of Latin America, about which he had plenty of ideas. One of his beliefs was that the very resistance the Latin-American continent seemed to offer to scientific exploration meant

153

that he, Humboldt, was the man most suited to go there and demonstrate from observing the local phenomena the interdependency of all the sciences. He did not believe in the curse of the Green Hell, a superstition he attributed to an ill-informed approach to a world undeniably full of enigmas. On botany, also, he had much to say, protesting strongly against those he dubbed "wretched archivists of Nature." As he said:

> General harmony of form, the problem of knowing whether there is a single original plant form, which is then manifested in thousands of gradations; the distribution of these forms across the Earth's surface; the various feelings of joy and melancholy aroused in sensitive men by the plant world; the contrast between the dead, motionless rock masses and even the seemingly inorganic trunks of trees and the living vegetable carpet which delicately clothes, so to speak, this skeleton with a more tender flesh; the history and geography of plants, that is, the historical description of the spread of vegetables over the earth's surface ... the investigation of the most ancient primitive vegetation in its funeral monuments, petrification, fossilization, coal; the progressive habitability of the globe; the migrations and movements of plants, as groups and in isolation, with maps showing what plants followed what peoples; a general history of agriculture; a comparison between cultivated plants and domestic animals ...; a study of what plants are more or less closely subject to the law of symmetrical form; the natural reversal of cultivated plants to the wild state ...; the general confusion of plant geography provoked by colonization; these, in my view, are subjects worthy of attention and as yet almost totally neglected.

This scholar was certainly not short of ideas. Among his listeners one young botanist, Aimé Bonpland, was particularly enthusiastic. Bonpland, another regular guest in Bougainville's salon, was a specialist in exotic flora. He could not fail to be intrigued by tales of the Amazonian jungle. He felt he could follow every turn of the mind of the young German, whom he considered as the most intelligent man of modern times, and that he could follow him to the end of the earth as well. In the gilt-paneled salon of Monsieur de Bougainville, returned from his own travels, Humboldt and Bonpland formed a strong lifelong friendship that eventually took them to the most uncertain paths of the world.

Aimé Bonpland was a quiet young man, twenty-five years old when he met Humboldt. Born in 1773 in the port of La Rochelle, he was the son of a surgeon who spent all his spare time gardening in the grounds of his house. A vine specialist, the father had several trial plots for his plants. Aimé's mother was the daughter of a ship's captain. The son appeared to combine traits of both his parents, for he decided to become a doctor but to practice in the navy. Once Bonpland arrived in Paris to study medicine, he fell hopelessly in love with the exotic flora in Thouin's glass houses. From then on this retiring, peaceful boy, was consumed with a single-minded passion for wild plants.

Jules Verne writes that Humboldt was to join Baudin's expedition but that he grew impatient during its preparation. He then went to Marseilles hoping to join Bonaparte's expedition to Egypt. Impatient again, he went to Spain with Bonpland to organize their own voyage. Since no official body was willing to sponsor a mission to Latin-America, Humbldt, who was extremely rich, decided to pay for the journey. The two botanists spent a year preparing their voyage as carefully as possible, spending most of their time with Thouin in the greenhouses of the Garden, and in the Muséum library, where they asked Desfontaines to bring out all the manuscripts having to do with the expeditions of Joseph de Jussieu, La Condamine, and Dombey. Bonpland was interested solely in the flora, but Humboldt sought to gather as wide a range of information as possible. He hoped to prove two important convictions. The first was that the immense vegetable richness of South America, and even many of the natural phenomena noted there, were to be explained by an unusually high level of magnetism in the area. The second was that South America was an important cradle of civilization which so far had been inadequately investigated. So long as the inscriptions engraved on the rocks and monuments remained undeciphered, Humboldt believed that attempts to explain this world would be vain. He intended to contribute to such an inquiry, in spite of the fact that the Spanish had so far prevented any archaeological work from being done in these countries.

The attitude of the Spanish authorities had been one of the worst problems encountered by Dombey. But now the situation was different, because Humboldt, a powerful member of the Prussian aristocracy, could not be lightly dismissed. More important was the fact that he was not working for any government but was undertaking his expedition as a private individual. Such an arrangement was less wounding to Spanish national pride and involved fewer economic risks. The Spanish could not overtly stand in his way. Humboldt was a skilled negotiator and knew how to obtain his wish. He dealt directly with the king of Spain and discussed with him every detail of the passports and authorizations he and Bonpland would need during their travels. "I obtained our passports," he relates, "and never had fuller permission been granted to travelers. No foreigner had ever been more trusted by the Spanish government. To dispel any doubts the viceroys or representatives of Spanish power in America might feel concerning the nature of my work, the passport indicated that I was permitted free use of my physics and geodesic instruments, and that I could make astronomical observations in all the Spanish possessions, measure the height of mountains, gather the produce of the earth, and conduct any operations I judged useful to the progress of science. Since the purpose of our journey was purely scientific, we succeeded, Mr. Bonpland and I, in winning both the goodwill of the colonials and that of the Europeans charged with administering those vast territories."

This was indeed a most generous concession from a government normally so jealous of its interests. Another reason for the generosity was the fact that Humboldt had made a firm commitment with regard to the eventual communication of whatever he might discover. He would forward a summary of his work to Germany, England, Spain, and France, and send all his botanical finds to the Muséum d'Histoire Naturelle and the Jardin des Plantes in Paris. This arrangement was not mere strategy on Humboldt's part: A believer in the unity of science, he also longed for the universal spread of knowledge. If he wished to share the outcome of his work with England, in no way involved in the expedition, it was out of gratitude to Sir Joseph Banks (the former botanist to Cook), who, as we have seen, was always seeking to aid his colleagues and had opened to Humboldt his exceptionally fine library. Humboldt had also gathered plants in Germany with the naturalist George Forster, who had accompanied Cook on his second voyage.

All their plans settled, Humboldt and Bonpland left the coast of Spain at La Coruna, on June 5, 1799, aboard the corvette *El Pizzaro*, and headed for Mexico by way of Havana. On the boat, various adjustments had been made to install Humboldt's instruments for he wanted to conduct meteorological studies and experiments on the chemical composition of air during the voyage. Since the crossing was a long one, the two men had time to dream about what awaited them and to continue to exchange ideas. Humboldt again stressed to Bonpland that, although he believed in exact sciences, he also remained convinced that the sciences constantly modified one another, that nothing was ever immutably fixed and static. This notion would always disturb, even frighten Bonpland, who was raised on classical theories about the autonomy of each discipline. He agreed, however, to reconsider the classification of flowers in botany in order to escape too rigid a system of analysis. He shared Humboldt's view that instead of endlessly dissecting dried flowers in a laboratory, giving them names, and pressing them in herbaria, one should observe them in their natural setting and thereby come to understand why and how plants grow in a given spot and what is their role in the universal distribution of natural riches.

The ease, confidence, and charm of his companion impressed Bonpland more each day, and so it was natural that he should let Humboldt fix their work schedule. Humboldt proposed that they divide their tasks into sections, the first being an account of the expedition, others covering sociology, zoology, comparative anatomy, astronomy, geology, and botany. In addition, Humboldt intended to study vanished civilizations, as well as the role of electricity in the atmosphere, electrical waves, magnetic currents, and meteorological phenomena in relation to the growth and behavior of specific plants. Such topics obviously broadened considerably the scope of their inquiry. Humboldt constantly told Bonpland that significant gaps in knowledge resulted from insufficient endeavors to focus varying methods of research on a given subject. He thought that scientists might soon be able to abandon the vague notion of "vital force," widespread at the time, if they could establish that everything proceeded from the combined action of material forces all known separately but never before seen as related. In a sudden lyrical moment he would refer to Goethe, evoking the "impressions of joy or melancholy which the plant world makes upon sensitive men," words that could not fail to appeal to the young Bonpland, in search of absolute and secret wonders.

They landed in Venezuela on July 17, 1799, at Cumana, a port east of Caracas. Immediately they rented a house, less for themselves than for Humboldt's precious instruments. His fortune had enabled him to buy the best available marine and astronomical chronometers, telescopes, sextants, dipping compasses, barometers, and thermometers, as well as firearms. Once installed, they ventured out, and were so astonished by what they saw that they could not study for several days. They felt they were on another planet, where the vegetation had tamed and enslaved the humans. Their excitement was indescribable. The first messages they sent back to Europe were that everything in this world was beyond belief, built on a scale unfit for Europeans, and that they would have to reconsider all their habits of thought and observation if they were not to be completely overwhelmed.

They ventured out, surrounded by giant banana trees and a mass of other trees covered with immense leaves. The more they looked, the less they recognized what they thought they knew. They had not foreseen or even imagined such a gigantism, such an unleashing of green density. What was a small coconut cutting in a greenhouse beside entire forests of sixty-foot-high palms with stately, beautiful red blooms looking like a scarlet mountaintop on which one could walk under the sky?

Bonpland did not know where to begin. He saw thousands of flowers about which he knew nothing, with outsize corollas of brutal colors as dazzling to the eye as to the mind. Day after day he discovered the meaning of excess and, setting feverishly to work, began his quest. Humboldt, hardly less overwhelmed, immediately understood the true nature of the problem: In such a world, in this luxuriant and sumptuous chaos, startling, poisonous, and unfathomable, what was the place of man, if he had any?

Soon the two friends ventured into the interior. As they advanced, they became daily more conscious of the truth of the wildest tales, the most terrifying as well as the most enthusiastic reports. At the same time, they realized that everything still remained to be said and done. And yet so much had already been said, ever since the Spanish conquest. The Spanish chroniclers had no words disquieting enough to describe the Green Hell. Pietro Martyre d'Anghiera asserted in 1494 that the Indians fed on human flesh. These natives, he added, "rejected the missionaries, shooting poisoned arrows at them." Fernandez de Oviedo, who stayed in Central America from 1550 to 1556, reported that he saw "jumping vipers" which attacked men, and twenty other kinds of deadly snakes. He told of spiders as big as birds, of enormous lizards, gazing out from cold eyes, of toads comparable to cats, and, above all, of the ceaseless howling of animals calling to one another in the forest, creating a kind of music of the spheres all the more fearsome because so bewitching. The Dominican Gaspar de Caravajal, about 1580, described how he and his companions had been encircled by hordes of militarily organized mosquitos and how it seemed impossible not to be eaten alive after having been driven mad with bites. During the same period the Jesuit José de Acoste, staying in Peru, near lake Titicaca, decided that the animals in this New World could in no way be God's creatures. Others wrote of soldier insects, of enormous bees with poisoned honey, and people scarcely dared mention the terrible armies of red ants. In 1707 a German Jesuit, Father Fritz, explained at length that it was impossible for men to penetrate the forest, because if, by chance, they managed to hack their way in, they later perished for lack of air, or from breathing a thick mist full of unbearable and lethal vapors.

But almost no one had spoken of the plants. The conquistadores had come for gold, and the missionaries for souls. Not until the French botanists arrived did the world discover rubber, quinquina, cocoa, cinnamon, mahogany, and even maize, which when first brought back by Columbus had not interested anyone. Not until the time of Joseph de Jussieu, La Condamine and then Dombey did Europeans at last marvel at the blue orchid the Indians called "the eyes of the earth goddess," and learn that the Green Hell also contained sumptuous floating gardens of multicolored wild hyacinths.

As if nature were not threatening enough, everytime they wanted to make camp, Bonpland and Humboldt found themselves face to face with naked men and women, their faces painted or masked in thick, cocoonlike objects, their bodies sometimes caked all over with dried mud. The scientists lived in constant fear of falling into the hands of the dreaded headhunters. Humboldt multiplied his observations; he noticed that the Indians were constantly chewing something, a plant, a piece of bark, or a berry, and that from these same ingredients they derived alcoholic beverages and drugs which they used for ritual purposes and for communicating with their gods. The two friends were finally led to study curare and to note the extreme skill with which the Indians used it. "I know," a medicine man told them, "that the Whites have the secret of making soap and that black powder which has the defect of making a noise and driving off animals when they are not hit. But the curare we prepare according to a recipe handed down from father to son is superior to anything you can make. It is the juice of a herb which 'kills silently.' " Curare did indeed have many uses, from hunting and warfare to calming digestive upsets. The scientists watched its preparation, and Humboldt conducted various experiments, some of them by accident. On one occasion the worst almost happened: "The curare, having attracted the damp, had become liquid," he wrote, "and had leaked out of a poorly closed jar onto our linen. When washing the linen, we forgot to examine the inside of one stocking, which was completely filled with curare. It was only upon touching this sticky substance with my fingers that I was warned not to put on the poisoned garment. The danger was all the greater because my toes were bleeding at the time from the bites of chiggers. This incident," Humboldt concluded calmly, "should remind us of how careful we must be in transporting poisons." Back in Europe, he was to publish a study on the poison which "kills silently."

Tirelessly Humboldt watched the sky, dominated there by Venus: "This planet," he wrote, "here plays the role of the moon. It has great luminous haloes, with the loveliest rainbow shades, even when the air is dry and the sky a clear blue." Bonpland, meantime, gathered all the plants he could. He described their nature at length, corrected older classifications, and divided the subcontinent into climatic zones. At the end of 1799 he sent to Paris, by way of Madrid, the finest collection ever received from Latin America. But although Bonpland never failed to send off his consignments, he did so with no great feeling of hope. After living in the jungle, he realized that rubbers, ferns, rhododendrons, and begonias of the Andes, the ficus, caladiums, heliotropes, and cacti could only become, in Europe, faint and ridiculously tame echoes of the wild and mighty spectacle unfolding before his eyes. This first consignment was extremely rich, for it included a hitherto unknown ornamental tree, the cashew (*Anacardium occidentale*), almost fifty species of passionflower (*Passiflora*), fuchsias, cineraries (*Senecio pulcher*), giant zinnias, and entire families of new orchids. In all, he sent over a hundred species of orchid accompanied by notes describing their habitat. Some, Bonpland explained, lived in a crack on a rock; others on charred tree trunks; and yet others on the lianas which enclosed the forest in a tangled mass, often entwined with passionflowers and bignonias. These flowers so hate each other, he observed, that they expend all their energy in seeking to destroy one another.

The two men worked constantly. Humboldt was fortunate enough to witness an earthquake, and he conducted some experiments on volcanic fires. On the walls of a grotto sheltering night birds which looked like flying toads, he found characters in an unknown writing. One day he studied wind and humidity levels; the next he observed various kinds of sugarcane. He and Bonpland first explored the more accessible regions just outside Caracas, where, wrote Bonpland, ferns grew as high as "our lindens and beeches," banana palms were gigantic, and grasses arborescent. One day, outside a village where they had visited cotton and indigo plantations, they saw a rare marvel. "It is," wrote Humboldt, "an object looking from a distance like a rounded hillock, a tumulus covered with vegetation. It is neither a hill nor a group of close-growing trees, but a single tree, the famous zamang of Guayre, known throughout the province for the huge spread of its branches, which form a dome of 576 feet in circumference. The zamang is a handsome species of mimosa: Its tortuous branches divide by way of bifurcation." Recovering from their surprise, they walked underneath, as if entering a cathedral of arching branches and flowers, where total silence reigned and the air was filled with a delicate but spicy perfume. The Indians accompanying them showed the most profound respect towards the tree and Bonpland was so moved that he almost fell to his knees. The botanist determined that this famous tree, a rain tree (*Inga saman*), locally called the zaman of Guayre, was several centuries old; the conquerers had seen it 300 years earlier, and it had existed well before that. It belonged to a group known as "civilization" trees, such as Adanson's baobab, Buddha's banyan (*Ficus benghalensis*) in the Ceylonese jungle, and the Tenerife dragon tree (*Pleomele draco* or *Dracaena draco*), said to be older than the pyramids when it was destroyed in 1868. They were considered to be the equivalent of monuments; those defacing them were severely punished.

The rivers, the great Amazonian mystery, still remained to be explored. The scientists' Spanish friends were alarmed when they learned of the men's intention to travel up the Orinoco River to northern Brazil to complete the work undertaken by La Condamine. These regions had hardly ever been explored, explained Bonpland and Humboldt; the Orinoco was fringed with forests where no white man had ever been. They set off in solid dugout canoes guided by Indian rowers. They did not realize that these rivers could run 4,000 miles and were sometimes several miles wide, so wide that one could not see the banks from the mainstream; they were unaware of the gigantic mass and incredible force of the water which moved like a running sea. Humboldt was determined to solve the mystery of these waters, and Bonpland that of the self-protected virgin forest. The trip lasted ten months, ten months of hell.

First of all, they encountered mirages caused by superimposed layers of air, a phenomenon Humboldt managed to describe and analyze. It was disconcerting to move along in a world in which animals appeared to walk upside down and in which the closest palm trees melted into a thin, hazy line on the horizon. Everything seemed to float; the rare sites without vegetation suddenly became huge lakes, their surfaces rippling.

Secondly, they were threatened by animals. The ferocious giant caimans and snakes had startling ways of attacking humans: "In summer," wrote Humboldt, "in the burning heat, the crocodile and the boa sleep motionless, buried in dried clay. Sometimes, on the edge of the swamps, one sees wet clay slowly rise up in patches. With a noise like the explosion of a small mud volcano, earth is thrown into the air. The spectator flees at the sight—a giant water serpent or armor-plated crocodile is emerging from his tomb, resurrected by the first rain." These animals could devour one in an instant and they were almost invulnerable inside their arrow-proof skin.

The electric eel, the gymnote, was scarcely less formidable, and was feared by the Indians more than any wild animal. Humboldt and Bonpland conducted various experiments on it, but emerged from their work exhausted, trembling, and aching, so strong were the shocks from the fish. In water it could kill horses and mules merely by rubbing against them as they crossed a ford. In a single day, the explorers saw seven horses perish in this way.

But nothing stopped them from working. Humboldt was convinced that the Orinoco and its banks had a prehistoric past, and so, while Bonpland collected plants, Humboldt hunted stones and engraved signs, witnesses of ancient times. Together, the two men studied Indian chemistry, for the Indians could do anything with plants, put their enemies to sleep or drive off mosquitoes. They crushed, distilled, measured, and mixed extracting drinks of every herb. Bonpland sent the Muséum vanilla and cocoa plants, cinnamon and palm trees, precious woods and more orchids, and mysterious plants to be analyzed in the Paris laboratories.

Humboldt, however, was worried. He continued to wonder exactly where and in what kind of world they were. "In the interior of this new continent," he wrote, "one almost grows accustomed to seeing man as not essential to the order of nature. The earth is overloaded with plants: Nothing impedes their free growth. An immense layer of leaf mold attests to the continuous action of the organic forests." Wild animals, "fearless and undisturbed, roam the forests, living there as in an ancient heritage. This animated nature, where man is nothing, is both strange and sad." As time went by, life in the Green Hell became even less pleasant. The days of torrid heat were trying, but when the humidity rose came illness and death. The forest dissolved into steam as thick as fog, all-enveloping beneath the leaden sky. The wild beasts became nervous and more threatening, just when dampness jammed their firearms: One of their horses was thus devoured by a boa. Bonpland found it hard to preserve his jungle flowers in these conditions; they seemed to droop and die instantly. The only creatures to thrive on the damp were the mosquitoes. The men were also troubled by nauseating exhalations from the water and fetid odors which rose from the undergrowth and befuddled the mind.

Others might have fled or given up, but Humboldt and Bonpland never seriously envisaged such a possibility. They could not leave these amazing floating gardens which, in the tangle of the river's tributaries, covered mile upon mile with what they were sure were hyacinths, water lilies, and fantastically colored heliotropes. But they had to stop for a while because, when approaching the Guiana Highlands, the exhausted Bonpland fell ill, consumed by fever and vomiting. Wondering if his companion had been bitten at night by an insect, Humboldt was unable to determine the nature of the illness. Their Indian companions tried various medicines, which perhaps helped keep Bonpland alive, but did not cure him. Humboldt remembered then how Bonpland used to discourse about the air: He had said how much he missed the pure, clean air of his native La Rochelle and how he felt uneasy in the stifling jungle atmosphere. "Since I saw that he would not recover in the city," Humboldt wrote to his brother Wilhelm, "I took him to the country house of my friend Don Felice Fareras, four miles from the Orinoco, in a higher, relatively cool valley. In tropical climates, there is no better remedy than a change of air. In this manner, my companion was restored to health in a few days. I cannot describe the anguish I felt during his illness: I would never have found another friend so loyal, active, and courageous. He has showed amazing resignation and energy during a journey filled with danger." Never sick, Humboldt did not come through the ordeal unscathed: He lost his hair, perhaps from fatigue. As soon as Bonpland's health was restored, they resumed their travels among forest, valleys, and volcanos. Every day was a victory over death. Yet Humboldt was jubilant, for he had discovered traces of an unknown civilization. Bonpland began a study of the many kinds of palms. Some produced fruits, delicious either raw or cooked, others a kind of wax, and yet others milk. One gave silk, and another bore 44,000 flowers. The Indians wore, ate, and drank palm products, to such an extent that Humboldt coined the expression "palmivorous man." Bonpland collected and commented on all these palms, sending specimens to Paris, where for a long time the tree had been popular both for its exotic associations and for its commercial possibilities. Meantime, Humboldt explored volcanic craters.

After Chile and Peru, they decided to go to Mexico, despite their exhaustion. They could hardly remember a time when they had not been living in flimsy tents or cabins improvised of leaves and branches, but their curiosity remained inexhaustible. They continued to study Indian chemistry and technology. Everything around them seemed to come from the vegetable world, and everything seemed to be reclaimed, sooner or later, by the forest. "Bonpland and I spent the night in an abandoned cabin," reported Humboldt. "The Indian family had left there fishing instruments, pottery, mats woven of palm petioles, the remains of a mixture of tree resins they had used to caulk their canoes, and bowls full of vegetable milk used for varnishing. With this viscous liquid, they coated furniture they wanted to color white." The natives also made varnish, and water bottles and roofs with rubber. Bonpland discovered a tree he christened, for want of a better name, "shirt tree" (*Astrocaryum mexicanum*), because the Indians used its red, fibrous bark to make long ponchos to wear in rainy weather. Humboldt concluded: "The more we study the vegetable chemistry of the torrid zone, the more we shall discover, half prepared in the organs of plants, products we thought so far to belong to the animal kingdom, or which we manufacture by artificial means always reliable, but often long and arduous."

Their discoveries were so numerous that it proved difficult to classify them satisfactorily. Therefore, they often limited themselves to dividing the specimens into broad categories, floating wood, colored woods, scented woods, woods for making boats or paper, and so on, sending their findings to Paris, Madrid, London, and Berlin for final classification. The result was, for the first time in history, an internationalization of research and results. The two men were appropriately rewarded by becoming famous all over the world. From the United States, Germany, and England, people came to visit at various stages of their journey, and left full of new information. These visits occurred more and more frequently after their Mexican expedition, when they had to give what amounted to press conferences. Academies and learned societies welcomed them with open arms; journalists left them no peace.

Humboldt was happy: The climate zones of Mexico proved exceptionally varied and interesting. He could observe rivers, splendid lakes, lake basins, mountains, volcanos, and deserts; he measured heat and cold, pressure systems and depressions, cyclones and anticyclones, Atlantic and Pacific winds. When he left his instruments, it was to study the ruins of pre-Columbian civilization.

Bonpland went from forest to savannah, valley to sierra, collecting all sorts of trees, guavas (*Psidium guajava*), new magnolias, numerous varieties of cactus, reeds for basketmaking, black poplars, juniper trees, agaves, tropical dahlias, tobacco, and maize. He sent to Paris the famous Mexican calla lilies, plants "of divine beauty," and sorted his herbaria, undertaking an immense, methodical task. Every evening, the two travelers met to discuss the events of the day. In Mexico they received an invitation from President Thomas Jefferson to visit him in Washington. Their expedition completed, they set off immediately for the American capital. On the way they stopped first in Philadelphia, where Humboldt was made a member of the American Philosophical Society and where they visited Charles Willson Peale and his museum. They arrived in Washington on June 1, and Jefferson honored them the next day with a state luncheon at his house. Jefferson was immediately taken with the explorers. As he had written to Humboldt just before he arrived, "The countries you have visited are of those least known and most interesting, and a lively desire will be felt generally to receive the information you will be able to give. No one will feel it more strongly than myself, because no one, perhaps, views this new world with more partial hopes of its exhibiting an ameliorated state of the human conditions." During his visit Humboldt stayed with the president and gave him valuable maps and statistical data on the territory America had just acquired by the Louisiana Purchase. All in all they stayed in America for two months, May and June, before leaving on June 30 for Paris. They arrived there in August 1804. Their expedition had lasted five years.

Paris in 1804 was a brand new empire: Bonaparte had become Napoleon I in May. Everything that might enhance imperial prestige was most favorably regarded. Since, next to warfare, science was the art most revered by the new Caesar, the welcome accorded the two travelers can be easily imagined. Humboldt was invited to join the French scientific institutions. He decided to remain in France to draft the first full account of his discoveries. According to recent calculations, Humboldt and Bonpland, in this one expedition, enriched the world's botanical treasury by 5 percent to 6 percent, having gathered during their five years in America almost 10 percent of the then known plants. Bonpland himself prepared and dried almost 60,000 plants belonging to some 6,200 species.

They first began work on their botanical writings, then on the others. Humboldt was to publish twenty-six volumes, ending his publishing life with his masterly *Kosmos*, the final expression of his ideas. Bonpland returned to the Muséum to work with Thouin, whom he greatly respected. But the trial plantings, acclimatizations, and routine seemed very far from Amazonia. The only good moments were those spent with Humboldt, to whom he constantly returned. Those five years they had spent in America they alone could understand and talk about. Not that other people did not urge them to talk. On the contrary, but to others they could only tell what must have seemed a fantastic tale. Still, they had to tell their story. Readers were tired of ships' logs, of endless travel accounts. Landings on unknown islands, first meetings with natives, strange customs and dress now seemed tedious. Even an account by Captain Cook would no longer excite. The public was looking for unusual adventures, accounts of individual moods, and subjective reactions to events. The two travelers were eminently qualified to satisfy these cravings. At first, Humboldt was shocked, but the two agreed to perform the task with some misgivings. "We agree," noted Humboldt, "with considerable reluctance to write an account of our voyage. I have even noticed that the liking for such compositions is so great that scholars, having presented their research separately, are not considered to have performed their duty to the public until they have described their itinerary." The *Voyage aux régions équinoxiales du nouveau continent* appeared in 1810. The work was such a success that it became one of the emperor's bedside books.

Once the book was written, Bonpland became daily more unsettled, more estranged from everything around him. The herbarium was in order, his notes classified. He could return to medicine but had no desire to do so. Unlike Humboldt, he had remained emotionally in Amazonia and he longed to return there. Someone, however, was to prevent him from leaving.

Humboldt and Bonpland knew the Empress Josephine well. She had received them on their return and, subsequently, they had sent numerous exotic plants to her. They had been dazzled by her gardens at La Malmaison, and amazed by her knowledge. She knew extremely well the flora of the West Indies, where she was born and raised, and also that of the Orient. The empress was an enlightened botanist, but passionate and despotic. La Malmaison was her domain, one of her reasons for living. She enlarged it continuously until she owned about five thousand acres in which she sought to recreate a tropical atmosphere.

Bonpland was fired with enthusiasm. Of course, this estate was not the great, green forest, but at least it contained the colors of his dreams. He decided to help the empress turn her grounds into a real botanical park with beds of mimosas, heliotropes, cassias, lobelias, cacti, and orchids. He reorganized the orangery, brought in new trees from Central America, Amazonian ferns from the Jardin des Plantes, and built several greenhouses. Bonpland and Josephine soon became close friends. She was beautiful, and she shared his liking for exotic flowers. She decided to create and run a nursery garden, and for this purpose she engaged Mirbel, a Muséum botanist. But Napoleon could not endure Mirbel, who ruled tyrannically over the garden, locked the glass houses, and induced Josephine to spend incredible sums of money. He was dismissed and replaced by Bonpland, who henceforth controlled the grounds and farms of the empress.

He was happy during this period in his life. The magnolias flourished, as did the flowering cacti. He took notes on everything, giving himself tirelessly to his job. He tended to the still world-famous rose garden. With Redouté, he undertook an immense work, the *Plantes de la Malmaison*. Since all Europe knew of his position with Napoleon and Josephine, he was consulted from every land. He maintained a voluminous correspondence with the greatest botanists of the day, one of his most faithful correspondents being the aging Sir Joseph Banks, who invited him, "as soon as events permit," to visit Kew Gardens. Bonpland went also to Berlin and Vienna, where the Schönbrunn gardens, so often mentioned by Humboldt, delighted him. Everywhere, he was given rare plants and unique specimens. At the request of the empress, eager to spread her treasures over a wider area, he contacted the first great nurserymen—Vilmorin, Perregaux, Noisette, and Cels. With them began a new era in plant acclimatization, which now entered an almost industrial stage. Exoticism was for sale. In England nurserymen were already equipping special sailing-ships for collecting live specimens.

Bonpland and the empress spent many hours together. At La Malmaison the orchids flourished in ever-increasing numbers; they became Josephine's passion. "To our delicate and tender plants," wrote an indignant historian, "she preferred the flowers of her childhood, coarser and more violent, flowers seemingly fed on meat, so intensely alive that they are too frightening to touch." For some years Bonpland, constantly at Josephine's side, shared Napoleon's private life when the emperor took time to relax a little between campaigns. The emperor was somewhat irritated by his wife's caprices and by the amount of money she spent; one day he and Bonpland quarreled about a mere maple. Unfortunately, in 1809, Napoleon repudiated the childless Josephine, and Bonpland was obliged to relinquish his vicarious flower empire at La Malmaison. Alone again, completely lost, almost forty years old, he married in secret. We know nothing about his wife, except that he did not get on well with her.

In desperation, he turned once again to Joseph Banks, whom he visited in London, accompanied by his wife. They spent a long time in England where Bonpland was treated with great respect. He never tired of watching Banks, a great botanist, at work. The new nurseries were models of their kind, employing 150 gardeners, an unheard-of number. At Kew Gardens, Bonpland did not know which to admire most, the eucalyptus, the magnolias, or the Peruvian araucarias. But the reverse was also true. The English so admired Bonpland's work that they asked him to draw up a catalogue of their rare plants. This period was only an interlude. In London he reflected carefully on his situation and he came to realize that he was unhappy in love and in himself and insufficiently absorbed by his work. Finally he admitted in 1816 that there was no alternative but to go back to South America. He had little money, though, and did not know where to seek help. Quite naturally, he eventually turned to Humboldt, at whose house he met Simon Bolivar, a thirty-three-year-old South American exile. This passionate, despairing young man had lost his wife and suffered deeply because of the state of oppression under which his fellow men were living in Venezuala, of course, but also in Mexico, and in what was later to become Panama, Colombia, and Ecuador. Having made himself the champion of their independence, he was preparing to return to Latin America. For the time being, however, he was a defeated man who could find sympathy only among scholars—Laplace, Gay-Lussac, Humboldt, and, now, Bonpland. This circle of friends longed for social justice and dreamed of adventure. Bonpland was not immune to such a climate of opinion. If Latin America was to begin a new stage in its history, he intended to contribute to the change in any way possible. He was familiar with these countries' agriculture, knew what they could grow and sell, and promised to advise the future independent nations. Soon he was able to tell Humboldt that he was about to set off, helped by Bolivar and his friends.

Humboldt feared that he would never see his friend again. For a moment he was tempted to go too, feeling that his own youth and passions also remained behind over there. But his country needed him, and his scientific work was far from complete. He could not abandon everything, after so much effort and with so much left to say. He resigned himself to worrying about Bonpland, giving him endless advice during the long farewell evening they spent together. He asked the botanist to write whenever he was in trouble, promising to send him help—a doctor, an explorer, an ambassador —wherever he was. Bonpland told Humboldt not to worry: He was familiar with the Green Hell and had returned safely from it. His marriage broken, he wanted to make a new life. He promised to write frequently and to send regular consignments of plants. Before Bonpland left, they together made an official presentation to the Muséum of their herbaria and set their papers in order. Humboldt needed some letters of recommendation for his friend. Bonpland was named correspondent to the Muséum, and he embarked late in 1816. Napoleon was on Saint Helena, Louis XVIII on the throne. Josephine had died at La Malmaison in 1814.

The age of the conquistadores was now past; that of the *libertadores* had just begun. Bonpland went first to visit one of these liberators of the new Latin America, the Argentinian Rivadavia, whom he had met in Paris at Humboldt's residence. After some initial difficulties, he settled in the province of Corrientes, in the far north of Argentina, close to Paraguay. There he set about rebuilding a former Jesuit college. He tilled the long-abandoned land, and a year after his arrival his estate was so profitable that it was cited throughout the country as a model. At the same time, he fulfilled his promises to the Muséum, sending back 800 new plants, in exchange for which he obtained specimens of European fruit trees and vegetables. In 1821 he was in a position to give Argentina its finest and richest plantations of *maté* (*Ilex paraguarensis*), which is used in Latin America to make tea. He recorded six species, recommending their acclimatization in France, and also sent some "long tobacco," the *pety cupu* of the Guarani tribe.

His estate was sumptuous, with alleys bordered by black laurels, cedars, mahogany, and palm trees. The European vegetables and fruits succeeded well, enriched and magnified by the fertile alluvial soil. Bonpland's only disappointment was to learn that in France it had not proved so easy to acclimatize the avocado, guava, and banana trees, or the pineapple and coffee plants he sent back so regularly. Nonetheless, he was content to ride around his plantations, free from irksome controls. To be sure, Rodriguez Francia, the Paraguayan dictator, was nearby, but Bonpland courteously kept him informed of his activities, and believed he had nothing to fear.

Francia, however, had just decided that Paraguay must have the monopoly of the *maté* trade. That just over the border a cunning European, moreover a liberal, should defy him was intolerable. He was on the other side of the border; but borders then, in those regions, were rather theoretical. Not naturally suspicious, Bonpland had never paid much attention to the dictator's spies, who were forever prowling about his land. In 1821 he was happy, foreseeing a large crop; he was also satisfied with the work done in his laboratory, where he was then concentrating on trying to analyze the local indigo (*Indigofera anil*). In an act of unabashed violence, a band of armed men invaded the plantation, massacred the Indians, and burned the buildings and crops. Bonpland received a saber wound on the head and was led in chains before Francia. He was to be kept chained for several months, then forcibly detained in the country for ten years, until 1831. He spent these years in abject poverty, clothed in rags, a pauper among paupers, in an isolated village, under constant surveillance. To pass the time and to make himself useful, he became a doctor, preparing decoctions, syrups, and medications to treat the Indians and peasants. He was content with his lot, for at least he was not shut up in one of Francia's prisons, where men often died mad, broken by torture. This special treatment was due in part to the fact that Bonpland was French and famous, and in part to letters that came from all over Europe asking for his release. Bolivar intervened, but to no avail. Humboldt alerted all the embassies. Paris sent an envoy, Richard Grandsire, bearing letters from Humboldt, Thouin, Cuvier, Desfontaines, and Chateaubriand, then minister of foreign affairs. England and Switzerland entered the struggle but were no more successful. Francia either understood not at all, or understood all too well. He believed that *maté* should belong to Paraguay and if the fate of an unfortunate botanist created such a stir, perhaps it was best to keep him a hostage. He could be a useful object in negotiations with the great powers.

But there were other possible reasons. "El Supremo," as Francia liked to style himself, was highly impressionable, and Bonpland's ascetic life in captivity, his dedication and humility, were for once successful in checking the dictator's repressive impulses. Initially, he had taken Bonpland for a naive gardener, then for a formidable opponent, a diabolical inventor. Now he was forced to come to terms with the extraordinary influence his prisoner exerted over the people he, El Supremo, sought only to oppress. It was an influence both profound and discreet; Bonpland cared, cured, and was universally beloved. He was in the process of discovering a tendency and a dimension that he had never before been conscious of in himself, that of the missionary. Such a character fascinated the aging, ailing Francia. The tyrant would gladly have placed himself in the care of this strange doctor, but for the fear that he might become the laughing stock of his people. Curiosity turning to anger, he suddenly decided to banish Bonpland, hoping to free himself from temptation and to break the bond between his people and this humble, powerful man.

Bonpland was free, but destitute, emaciated, and, although only fifty-eight, aged. He returned to Argentina, from where he sent off his first letter to Humboldt. He said that he was all right and ready to make a fresh start. He remarried, this time to a South American, and he made two more attempts at planting *maté* and indigo. He was finally able, with the aid of the Argentinian government, to begin again near Corrientes, where he set up, with friends' help, a plantation larger and more beautiful than the earlier ones. He began again to send packages to the Muséum by way of the

French consul in Buenos Aires. Germany, England, and Holland wrote to ask him for seeds and advice. Brazil consulted him, and invited him to give a new impetus to that country's agriculture. The six years from 1837 to 1843 were rewarding ones: He had three children: a daughter, Carmen, and two sons, Amadeo and Anastasio. He also had 2,000 orange trees and enough *maté* to destroy anyone's hopes for a monopoly. He maintained a voluminous correspondence, sending off package after package all over the world. From the Muséum, scientists would write: "Thanks to all your directions, we will be able to make use of the barks, roots, and other substances you sent for experimental use in the arts and in medicine. We would also like to thank you, sir, for the numerous shipments of seeds you have sent. A rather large quantity of the plants grown from these seeds are now flourishing in our greenhouses, which we hope you will continue to enrich with whatever you are able to collect. If you could also send specimens of the cactus genus and bulbs or tubers of orchids and liliaceae, you would be making a precious contribution to our garden." Could a botanist hope for greater satisfaction and recognition? Bonpland sent off the orchids, bulbs, and cacti, which the French had been very fond of since Dombey.

At the same time, he was delving ever deeper into the mysteries of Indian medicine. He was not a chemist, but, initiated perhaps by the Indians, he managed to conduct his own analyses of mysterious substances. He sent to Paris all sorts of berries, roots, and leaves, to which he gave only their local names because they belonged to genera for the most part totally unknown in Europe. He made a selection of barks containing tannin, and sent back the bark of a tree the natives called *chuchuhuaasha*. In the twentieth century Indians told the American explorer Leonard Clark that this bark cured cancer. He also dispatched a giant castor-oil plant (*Ricinus*), and, among flowers, fresh seeds of a splendid member of the nymphaea family, the royal water lily, which he gathered in June in fresh-water lagoons. The Englishman John Lindley named it *Victoria regia*. As soon as it arrived, it was seized upon by horticulturalists, Gorce, in England, and Vilmorin, in France, who distributed it widely.

Humboldt began now to urge him to return while there was still time, to come back to publish his work. But Bonpland stayed. He founded a Museum of Natural Science in Corrientes, and made important donations to the Buenos Aires Institute. Oblivious to the passage of time, he rode around his grounds, surrounded by his disciples. For a moment, he even dreamed of opening a school of botany. Humboldt wrote to him again, but Bonpland replied, "Dear Humboldt, you know very well that the plantation must live." But how did Bonpland himself survive? His wife was dead: His two sons, half Indian and quite uncontrollable, apparently spent all day with the gauchos, riding the estate thoroughbreds. It was said that his daughter was beautiful, and rather nonchalant. In 1855 a French traveler brought back the news that the eighty-two-year-old Bonpland, on his still active plantation, was now leading the life of an ascetic, retired in two small unkept rooms. His children had no interest in his work or perhaps in him. Nevertheless, he continued to go out a good deal. Even at his age he would ride off alone to meditate in some dense and dark forest along the Paraguay frontier. He was there at peace in his green continent, every species of which he had managed to grow. The little movement of the animals in the bush, the concerts of howler monkeys, the flights of bats soothed him. Orchids and the mysterious humus in the undergrowth were somehow sustaining. Then he would return quietly to shut himself up in his two cells.

His life had been devoted to the great trees and to flowers, and marked by fear and fascination. He had been warmed by a few friends; there had been Humboldt and the lovely Josephine, leaning over those "flowers made of flesh." History had gone by—the monarchy, the republic, the empire, the restoration, the second empire, all seemed remote and far away. Nonetheless, Bonpland still remained in touch with everything and everybody through the consul in Buenos Aires and students returning from Europe. He also had a friend, Father Joao Pedro Gay, the parish priest, who helped him classify the last herbaria, which the priest had instructions to send to the Muséum in Paris upon Bonpland's death.

The faithful Humboldt watched over him from afar, sending von Bulich, Prussian minister for the Plata states, to ask for news. The minister's report must not have been reassuring, for Humboldt next sent Dr. Avé-Lallemand, a Breslau doctor, who was traveling in Argentina at the time. In April 1858 the doctor reached the plantation. It was still an impressive sight, but the lianas were already taking over, choking the roses, alleys, and even the walls of the buildings. When Dr. Lallemand entered the house, he found a sick, feverish old man, rising with difficulty from his campbed, but still smiling, peaceful, almost radiant. Talkative also, muddling dates, he mixed together and superimposed upon one another the various moments of his life, endlessly returning to thoughts of his ten-year detention in Paraquay. To these years, perhaps, he owed his continuing spiritual strength, which enabled him to talk at length to his visitor about his latest work, about his rose garden where he had produced so many hybrids, or about the coffee plant, the *maté*, and the fig, about which he was busy writing.

When he offered to show his gardens to the doctor, the latter met with a surprise. For Bonpland did not see the grounds as they really were, ruined and dismal, but as they had been when at their luxuriant best. "Look," he said, "over there." There was no longer anything to see, the jungle having regained control. But Bonpland had already passed to another topic and was now discussing Humboldt, their discovery of America, and their expedition. He had kept up to date with his friend's work, which was now studied in Argentina. He remembered absolutely everything, remarking that he sometimes saw the Orinoco caimans in his dreams. He recalled that

"Humboldt saved his life because of his theories about the importance of the air we breathe." Was Humboldt happy, his friend asked? Honored and admired, certainly, replied the doctor. Bonpland wanted to write to Humboldt, but his hand trembled too much. The doctor would have to describe everything to Humboldt. "Tell him about my trees, and don't forget that, with coffee bushes, greenhouse trials can be totally negative. In changing light, or in a draught, they won't flower." Avé-Lallemand promised not to forget anything.

For the moment, however, he was struck by the dilapidated state of the two rooms the botanist lived in, full of dismantled furniture and scattered manuscripts which the visitor quietly placed back on the desk. Bonpland rambled on, making plans for the future. His beautiful daughter Carmen entered to put two pewter plates and a piece of grilled meat on the table. There was no cutlery. Avé-Lallemand was horrified, but Bonpland, busy talking about yuccas, did not even notice. "Tell Humboldt I have succeeded: The yucca is a splendid plant when it wants to be. Propagate by cuttings, not by seeds. We need to make more use of cuttings." The doctor wanted to examine Bonpland and to take him to a hospital in the city, but the botanist refused, declaring that he was perfectly well. They said farewell. Bonpland remained on the doorstep, smiling: "Oh, and tell Humboldt I've sent him some crates of fossil shells which tie in closely with his research."

Bonpland, now eighty-five, knew the end was near. He summoned Father Gay, and with him carried out final classifications. His last herbarium with 3,000 new plants was to go to the Muséum in Paris; there were gifts for the Corrientes Museum and scientific and commercial treatises for the Scientific Institute of Buenos Aires. The latter were almost exclusively concerned with the cultivation of so-called wild plants. On May 11, 1858, his children and Father Gay accompanied the unconscious man to the hospital in the nearby town. He slipped from unconsciousness to death.

But now a strange event occurred. It was very hot, but in view of the veneration in which Bonpland was held in the region, there could be no question of a speedy burial. The old ascetic had to be honored with a splendid funeral. The authorities decided to have the body embalmed and display it for a few days. His children returned to the plantation, and Father Gay was absent for a few minutes, leaving the body near an open french door. At that moment an Indian rode up and stopped his horse in front of the small hospital. He was one of those who adored Bonpland; he was also drunk. He saw the scholar, of whose death he was still unaware, and whom he believed to be dozing. "Buenas tardes, señor Bonpland." Smiling, he doffed his hat, saluted once, twice, a third time. Finally he grew angry, came closer, drew a knife, and stabbed the body several times. This done, he remounted his horse and rode off, still drunk.

The effects of the embalming were destroyed, and the body could not now be transported. Father Gay decided upon an immediate burial, simple and anonymous, in the little local Indian cemetery, an unremarkable site but at least one surrounded by trees.

When Humboldt learned of his death, he had just completed his great book *Kosmos*. He himself was to die the following year, 1859, at the age of ninety. Both these men had respected the Indians, who in turn had revealed to them secrets withheld from the Spanish. This was an important lesson to learn at the dawn of the industrial era, when plants would soon be no more than "raw material."

The empress Josephine in her garden at La Malmaison. From a Sèvres plate. (Photo by Malmaison–E. Tweedy)

In the Amazon jungle: standing, *Humboldt;* sitting, *Bonpland.*
(Photo by Roger-Viollet)

[14]

The Last Dream:

Jacquemont

You're going to think I've become a complete herbivore. No, dear friend, one doesn't become a plant maniac at twenty-four. In twenty years, if I have some spare time, we'll see. Even then, I'll only be forty-four, and a forty-four-year-old man still has better things to do than botany. For the time being, it's just a pleasant pastime." The man writing these lines spent ten hours a day on his scientific work. Two years later, driven by the mania he denied having, he set out for India. He did not live to be forty-four.

There were elements in him of René, Werther, and Byron. Victor Jacquemont was born in Paris on August 8, 1801. His eccentric father, Venceslas Jacquemont, had once been a priest, but after abandoning the priesthood had joined the leaders of the Girondin party at the beginning of the Revolution. He had escaped the Terror by fleeing to Hesdin, his birthplace, in the north of France. Then he became a member of the Institut de France and an opponent of Bonaparte. This opposition forced him into retirement during the empire. He returned to Paris only after the Bourbon Restoration, when he became a senior civil servant.

His son, Victor, was the epitome of the romantic scholar. We do not know much about him as a child, except that in 1815 his father registered him to take courses at the Collège de France. An accident in a chemistry class forced him to leave to recuperate in the country, at the château of La Grange-Bléneau, in Seine-et-Marne, the home of La Fayette, an old friend of his father. There he became engrossed with botany: "Since then," he wrote later, "when I found, on the shores of the Mediterranean, myrtles and other plants from our orangeries, or, on the Lozère, trod for the first time on scented mountain fields, I also felt great pleasure. But I never again experienced the wild joy of my early, quite commonplace discoveries in the woods of La Grange."

Some years later, in 1820 and 1821, he followed Desfontaines's botany courses at the Muséum, added to his herbarium, and began to be interested in mineralogy. At the same time, with some botanist friends—Adrien de Jussieu, Andolph Brongniart, Kunth, and Hippolyte Jaubert—he founded the Société d'histoire naturelle of Paris. A trip to Cantal and Provence, and another excursion, with Jaubert, to Dauphiné and the Alps, enriched his herbaria and increased his knowledge.

Victor fought, however, to keep his independence of mind and freedom of judgment. "The world," he remarked, "I mean the salon world, which is not educated, but is not devoid of tact and subtlety of mind, considers botany as a pastime, and botanists as creatures harmless to society, but somewhat foolish and simple. The world is right: I believe that nine out of ten botanists fully deserve such a judgment. Consider how many, after some years, become no more than machines for remembering names, reading catalogues, and writing labels. The same is true of mineralogists, both those who can name and those who own all the 1,347 varieties of calcium carbonate." The term "harmless" botanists, recalling Humboldt's "wretched archivists of nature," was typical of the language of the burgeoning romantic movement. Mental activity was viewed as worthless if it was not "harmful," if it did not involve a gamble. Jacquemont was a serious gambler. He chose his game with care; then he allowed it to govern and direct his life.

He was introduced to the salon world and spent his free time there, particularly in the salon of the comte de Tracy, where he became friendly with Stendhal and Mérimée and fell in love with the singer Schiasetti. They went for long horseback rides together, and Victor read the new poets to her. Neither love nor society lessened his ability to work ten hours a day or to publish in scientific journals. In October 1826 he received his diploma as a corresponding member of the Lyceum of Natural History of New York. On October 29 he left for the United States, where he remained for a year plant gathering with a friend, Dr. Stevenson. Shortly after his return to Paris, the Muséum offered him the chance to leave for India, because the French, unlike the English, did not have much information about this area. Chosen because he was both available and extremely competent, Jacquemont accepted eagerly. He wanted to escape from catalogues and labels. The departure date was set for August 1828, so he had almost a year to prepare. He reread Humboldt's early works on general physics and geology, plant geography, and comparative climatology. He had the greatest admiration for the German scholar, about whom he wrote: "Humboldt is one of those cultivating the sciences who are most perceptive in discovering new relationships, which is always important, between known facts." These "new relationships" between facts were what interested Jacquemont.

He set off, leaving behind Stendhal, Mérimée, Rossini, and the lovely diva who had caused him so much suffering. His boat, the *Zélée*, sailed from Brest on August 13, 1828, and stopped at Santa Cruz de Tenerife, Rio de Janeiro, and the Cape of Good Hope, where he met Dumont d'Urville, who was on his way home from Polynesia with sizable collections. On April 11, 1829, he reached Pondicherry at last. Then, on May 5, he was received in Calcutta by Lord William Pearson, solicitor general of Bengal. Victor set to work immediately: He learned Hindustani, pored over the books in the town library, botanized in the countryside, and studied the rich collections in the herbarium and botanical garden of Calcutta.

He also discovered the society of British India. Thanks to his letters of introduction and, above all, to his charm and distinction, he soon became the darling of the colony. He observed the British colonials without indulgence, alarmed to see how cut off they were from the reality of the country. He was the guest of all, particularly of Lord William Cavendish Bentinck, governor general of India. The British became so attached to him that they did all they could to keep him from leaving. Why did he want to travel around the country when it would be so simple to send out collectors to bring back anything he might want? It would be so pleasant to keep the handsome charmer in Calcutta, to chat about letters, the arts, even science, if absolutely necessary. He could supply the ladies with details about the latest Paris fashions! Jacquemont merely smiled his thanks. He was determined to become familiar with every aspect of India. "I am steeping myself in India, instead of just dipping my finger in it, as do so many English claiming to study it," he wrote to his father. He was not there for salon life or for embassy parties.

On November 20 he left Calcutta at the head of a small caravan composed of two ox-drawn carts and eight servants. At first he followed the Ganges plain, hoping to reach as soon as possible the western part of the Himalayas, which he intended to explore systematically. On December 31 he was in Benares, on March 5, 1830, in Delhi, and on April 2 in Saharanpur. The journey had not been particularly interesting. His journal emphasizes the uniformity of the plains, their monotony and lack of beauty. A note of disillusionment can sometimes be heard: "I was very disappointed when, instead of tigers and rhinoceroses, I saw only naked, unarmed men, calmly cutting wood to load onto carts. The forest resembled a wood thinned out a few years previously, but with several saplings left standing. It was possible, though awkward, to ride through it on horseback. The trees were all the same size and of the same species." He gathered only just over two hundred plants, but he made numerous observations concerning the botanical and agrarian landscapes, with notes on agricultural techiques. Wherever he went, he observed methods of cultivation and useful plants, as well as the natural cycles. "The cycles of vegetable growth," he remarked, "do not occur together. Each season has its own flora. The yellowing flowers of many terebinthaceae, leguminosae, and rubiaceae, which become almost bare in winter, fall onto a forest of shrubs ready to bloom. Meantime, yet others are ripening their fruits, held back during the rains." He did see some tigers at last, when, around Delhi, he took part in a week-long hunt.

On April 11, 1830, he was at the foot of the Himalayas. "When the sun began to set, I crossed the torrent and reached for the first time the foot of these famous mountains." Having crossed his Rubicon, Victor felt pride and confidence, but also anxiety. He knew that soon he would be completely cut off on these mountains, the most fearsome in the world. Soon he began the difficult walk over blizzard-swept peaks, punctuated by nights in tents exposed to the bitter cold of this land of extremes. He reorganized his caravan: The baggage was now to be carried by the servants, and he himself was to walk. He began to experience considerable pain from the first attacks of the illness that would one day kill him.

This first visit to the Himalayas lasted until November 16, when he returned to Saharanpur. The expedition had been extremely worthwhile. It yielded 2,158 plants. Jacquemont had seen many trees unknown to him and discovered, among other things, a new palm, later called *Phoenix humilis*, so small that it was "scarcely taller than the grass." He also made two excursions into China, and even fell down a ravine. He had observed the cultivation of the opium poppy: "I almost forgot to describe the growing of the poppy. Every family had its small field, either to eat the seeds or to make a little opium, which the men mix with the horrible ingredients they smoke. Each man also grows a few tobacco plants, the leaves of which are then mixed with the leaves of hemp, which grows wild all round the villages."

The long mountain climbs had been exhausting. Some of them lasted for twelve hours at a time. Tirelessly, Jacquemont had scaled each peak. He described each one in his journal, confirming, completing, and modifying what he had said the previous day. The inevitable repetition was for him unimportant beside the need to learn to distinguish and discriminate, to recognize the *same* and the *different,* despite often deceptive exteriors.

India was a strange country, violent and contradictory. Full of some of the finest rhododendrons in the world, it also contained familiar trees, oaks and walnuts, apples and pears. But tropical forms predominated, barberries (*Berberis* of various species, mostly *asiaticae, tinctoriae, angulosae,* and *sinenses*), growing beside pomegranates (*Punica granatum*) and wild vines. In the valleys, he found banana and apricot trees: The dried fruit of the latter "forms an important part of the local inhabitants' diet." Maples also grew there, as did the dazzling white peony (*Paeonia albiflora,* recently renamed *P. Lactiflora*), unknown before Jacquemont, but today the ancestor of most of our finest hybrids. Jacquemont had also approached that "region of eternal snow, where all vegetation ceases," except for a mysterious citruslike tree.

At the end of 1830 Jacquemont was back in Delhi, where, on December 31, the officers of the province held a banquet in his honor. He began to prepare at once for his trip to Punjab and Kashmir, a journey which lasted ten months and from which he returned with 1,500 plants, many of them new. This time, he also saw many tigers, which "live in a region with a climate very like that of central France. The tigers carry off sheep, attack horses and horned animals, even men, on occasion, when they are hungry during the winter."

Jacquemont was sick again. He suffered from a chronic colitis which would attack, subside, and then return. Convinced that he had to follow a strict diet, he set about leading an almost ascetic life. "My meal is simple," he wrote his father. "I take a large cup of cow or buffalo milk, or goat's milk, if need be, and some coarse-ground wheat pancakes." He had many ups and downs: "Six days on foot and on horseback have completely cured me.... Like a true Muslim, I have vowed total abstinence from spirituous liquors. I live much like the natives, finding, after many experiments, that this regime suits me best." He later wrote to his father that this life brought him "a strange sense of inner satisfaction," and added that "on considering my new state of mind, I feel a kind of savage pride."

The problems of his health led him to become interested in those of others. The death rate was high in the English colony: Jacquemont observed that "falls from horseback figure immediately after chronic hepatitis and ahead of cholera on the list of causes of death in this country." In his attitude to illness, that of others as well as his own, even when his sufferings were most acute, we can best appreciate the truth of Mérimée's claim that Jacquemont, for him, "best represented the Greek Stoic ideal, but in an affable manner, full of grace and gaiety." Jacquemont's stoicism was characterized by distance and humor. "Cholera," he wrote, "is ravaging the region of Mewar through which I traveled recently, but it only attacks the Indians. They say that water-drinkers are more susceptible to the disease than others. I'm at least going to tint my water red."

From medical observation, Jacquemont soon progressed to the practice of medicine. He received from Lord Cavendish Bentinck the title of *seigneur médecin*. This change was brought about by several factors—a sick man's fascination with illness, a real desire to help others, and the prospect of the contacts the practice of medicine could procure in a colonial and semifeudal society. He cared for all, rich and poor, from the rajah Ranjit Singh, who called him the "Plato and Hippocrates of our century," to the peasants. "The number of sick people coming to see me is endless," he wrote. "A crowd of poor and sick often presses about my tent, as a happier crowd does around our theaters. Unfortunately, most are incurable —blind people of all sorts, wretched souls eaten away by dreadful diseases caught from us. I give alms to those I cannot help medically, thinking sometimes with pleasure that at least a few sufferers go away with a feeling of gratitude."

But Victor was not really moving toward sainthood. When opportunities for pleasure occurred, he did not refuse them. The following incident, reminiscent of a musical comedy, took place in Lahore in March 1831. "Our compatriots gave me a most elegant party in my own palace, complete with charming Kashmir dancers. One of these girls would have passed for pretty, even beautiful, in any country in the world. I don't know how it happened, but at dusk, as the servants lit up the room, I found myself alone with this stage princess. The others had cunningly withdrawn into the garden with the rest of the group. Cunningly, and kindly."

In Kashmir, Victor's collection of plants was enriched with more than six hundred items. He gathered, among others, a currant bush (*Ribes alpestre*), some nettle trees (*Celtis sinensis*), wild pears (*Pyrus indicus*), a service tree with fruits the size of an apple (*Sorbus nepalensis* and *hupehensis*), alders (*Alnus nitida*), euonymus, azaleas, roses, jasmines, anemones, aquilegias, euphorbias, and quantities of rhododendrons.

He set out for Bombay on February 14, 1832. The country was covered in "green as far as the horizon, and completely empty of trees." He continued to gather plants and to work on his catalogue and journal. His correspondence became frenetic. Writing for him was as natural as breathing. He spent the summer in Poona, in the Deccan, where he packaged up his collections. Late in August he learned that King Louis-Philippe had made him a *chevalier* of the *Légion d'honneur*; by then he was suffering from dysentery. "For three days, I have been very unwell," he wrote to a cousin. "The pain was atrocious, but my head, completely clear, felt strangely fresh and alert. The activity of my thoughts was exhausting. I had the feeling that beautiful tunes by Mozart played on a fine violin would soothe my suffering. And so, since there happened to be an above-average musician here, I planned to send for him, intending to die with music. But then the remedies I had been taking finally brought about a reaction, and set me on the road to recovery. . . . I am now fully recovered; I am indeed better than I was before this attack. The illness was endemic. Cholera has killed many people here, but we are so used to it that we think no more about dying than a sailor at sea does of his ship overturning."

But he had not, in fact, recovered; he was growing worse. He wanted to go to Madras, a long trip, or at least to Bombay, but he was halted en route by his illness. "You can't fool with your stomach in this damned country. I covered myself all over with leeches, making it very uncomfortable to ride a horse afterward, but still could not completely dislodge the enemy. Nevertheless, I shall go to Bombay the day after tomorrow. The sea air I'll breathe in from the balconies there will do me good. The devil take the sun."

The enemy was not to be dislodged. The letter just quoted was dated October 27, 1832. On the 29th, Jacquemont did leave for Bombay, where he was immediately hospitalized. On November 4 he made his will, and on the 6th he was found to have a liver abcess. His condition worsened daily until on December 7 when he made final arrangements to send his manuscripts and collections to France and to organize his own funeral. That done, he died, at the age of thirty-one. He was buried at the cemetery of Lahore, in a bare tomb beneath a giant palm.

Jacquemont had been one of the last adventurers; but he was truly a botanist. After his death, his older brother, Porphyre, was charged with delivering all his documents to the proper scientific authorities. These consisted mainly of a manuscript catalogue of the plants he gathered during his journeys, presented in notebooks grouped into three sections: from Calcutta to Delhi, together with the Himalayas, Punjab, and Kashmir; from Delhi to Bombay; and a special notebook devoted to the Calcutta botanical gardens. An important feature, rare at the time, was that with each plant he provided a description of the characteristic localities where it grew and their ecological conditions. All this listed about 4,700 plants in all, an impressive number. The collections themselves did not reach Paris until November 1833. Later, they were incorporated into the general herbarium of the Muséum, where they are still housed. Duplicates were distributed to other important European botanical establishments.

After his death Jacquemont became famous as a writer. The publication in December 1833 of his *Correspondence* took the scientific and literary worlds by storm. It was the first modern picture of India, with an accuracy, depth, and breadth so impressive that even the English, not very favorably portrayed by Jacquemont, were full of admiration. In 1834 the *Foreign Quarterly Review* of London called Jacquemont's account "the most impartial and accurate ever written by a European." The appearance, in eighty instalments, between 1835 and 1841, of his *Voyage dans l'Inde*, his journal, confirmed this earlier judgment.

Humboldt studying "air" volcanoes. (Paris, Bibliothèque Nationale. Photo by ERL)

Victor Jacquemont. From a portrait by Mérimée's mother.
(Paris, Bibliothèque Nationale. Photo by ERL)

[15]

Aristotle Was Right

I T is not easy to conclude this story. In the late nineteenth and early twentieth centuries, industrialization began to control man's fate. Colonization and new means of travel modified space, photography and cinema the representation made of it. The conquered lands were carefully checked, scientists working together with the rest of the team. Emphasis was placed on the useful. Classification and our knowledge of plants gained in the process, but the adventure of travel had almost disappeared.

Was Joseph de Jussieu really looking for something, or trying to find or lose himself? How can we fail to think, even though there is nothing on paper to prove it, that Victor Jacquemont sought the death he found in India? In the lives and deaths of these wanderers there are strange constants, due, surely, to the fascination of flowers. By the end of the nineteenth century, no one went off to die for flowers anymore. Almost everything was available in botanical gardens and in the big nurseries. Now the task was to deepen the knowledge of plants, to order, to revise classifications, and to try to systematize, a difficult task for botanists constantly overwhelmed with material from all quarters. Laboratory and museum work was enormous; the field of study was almost limitless. Everything possible had already been acclimatized. Scientists began to study the anatomy, chemistry, and social life of plants. This age was also one of pharmaceutical discovery and of the use of plants as raw material.

Botanists influenced national economies by introducing and transplanting cane and coffee plants from one country to another. Exchanges between countries took place. Kew Gardens and the British Museum in London became the favored correspondents of the Jardin des Plantes and the Muséum d'histoire naturelle in Paris. The two colonial empires had accumulated gigantic herbaria.

New sciences appeared—taxonomy, the science of the laws of classification; cytology, the study of living cells; biochemistry; and embryology. New terms were invented—*pollen trajectory, alkaloids*, and the like. An exhibition in 1906 organized by Vilmorin, the French nurseryman, presented for the first time a fabulous collection of flowers, all of them from China and Asia. The show, accompanied by a sale catalogue, was a resounding success. The flower show soon became a familiar social event. Nurserymen became specialists, some known for apartment and garden flowers, others for cacti. Still others refused to sell anything but orchids. Plant magazines were published and societies of amateur botanists were formed to circulate ideas and suggestions. They brought into being a new art of gardening.

"We observe in plants a continuous ascent toward animal life," noted Aristotle, who, though he denied that plants could feel anything, believed they had a soul. A soul devoid of sensations was a strange concept, but it was accepted in the Middle Ages. In the Renaissance, the idea was swept aside as one of many myths. Flowers were acknowledged to have medicinal properties, a virtue that was only permissible when it did not savor too much of witchcraft. The eighteenth century, however, although just as rational as the sixteenth century, produced Linnaeus, the father of modern botany. He declared that plants were distinguished from men only in their absence of movement. This assertion was contradicted a century later by Darwin, who demonstrated that plants were capable of movement, but they "only acquire and manifest this faculty when it can be of some use to them." At the beginning of the twentieth century, Raoul Francé, a Viennese biologist, found that plants moved on their own account, as freely and harmoniously as animals or men, but infinitely more slowly.

The traveling botanists had always sought, in their accounts, to draw attention to the unknown life of plants. They could only send them back in the form of seeds and cuttings, or pressed between the pages of herbaria, but they had seen them in their natural state. They described flowers that close up as a passer-by approaches to hide the beauty of their corollas, the "drapes around the bridal bed," as Linnaeus called them. They told of creepers that suddenly cling to the branch or trunk of a tree, only to reappear once the danger was past, and of minute cacti that bury themselves in the sand at the slightest unfamiliar sound.

They could not believe that the vegetable world was devoid of purpose and consciousness, once they learned that the petals of a certain orchid imitated a female fly with such skill that males of that species sought to mate with the petals and thus pollinate the flower. It is surely not by chance that the marvel of Peru (*Mirabilis jalapa*) is white, given that we now know that the plant's reproductive process requires that color to attract moths. Today we realize that the only plants to have simple, straightforward beauty are those that rely on the wind for pollination.

Humboldt, Jacquemont, and many others knew about these complex phenomena. Their heirs attempted to find explanations for them. Men discovered sentinel plants and telepathic plants, which respond to all about them, near or far away. The travelers in the Green Hell would probably not have been unduly surprised to learn of such research. A botanist who paid the greatest attention to the unsuspected faculties of the vegetable world, at the end of the last century, was the American Luther Burbank. Born in 1849 in Massachusetts, he was impressed during his youth by reading

Humboldt and Darwin. His early botanical endeavors were grafting and crossing plants to produce new fruits and flowers. He must
have had a way with plants, for soon he had only to walk along
the paths of his garden for gladioli, zinnias, and clematis to grow
more quickly than usual. Burbank attributed many powers to
plants: "The most obstinate thing in the world," he wrote to a
scholar friend, "the hardest to handle, is a plant fixed in its ways.
Remember that such a plant goes back millenariums, as we know
from fossils. Don't you think that after all these repetitions, the
plant may have acquired a will of its own, if you want to use such
a term, an unbelievable tenacity?" Then there is the case of the
cactus to which Burbank spoke gently and at length: "You don't
need thorns to defend yourself," he would tell it. "I will protect
you." That cactus grew with no thorns.

Burbank lived in San Francisco during the famous 1906 earthquake and fire that destroyed most of the city. His glass houses
miraculously survived. At the end of his life, in 1929, he wrote:
"Plants have 20 different sense perceptions, but, because these dif-
fer from our own, we do not recognize them." He was not sure that
flowers actually understood his words, but he had no doubt that
some form of telepathy enabled them to discern his meaning.

A recent book, *The Secret Life of Plants*, by Peter Tompkins
and Christopher Bird, casts some light on the powers of the vegetable world. The authors recount the experience of Clive Backster,
the famous specialist in lie detection. In New York in 1966 Backster, in his office, which was filled with tropical plants, glanced at
a Dracaena, a tropical plant of which he was particularly fond. He
thought of connecting one of the electrodes of his polygraph (the
machine he used to detect lies) to one of its leaves. Immediately,
perhaps because the plant felt threatened, the needle oscillated.

The dialogue between man and plants took on a new dimension at that moment. Like a cardiogram or an encephalogram, the
machine proved that plants have an emotional rhythm, and that

they react to external events. Backster continued his experiments, and it became increasingly apparent that plants vibrated in response to unformulated thoughts, ideas merely passing through his mind, as well as to aggressive or friendly gestures, even before he tested them.

His experiments aroused wide interest. They were repeated by professional botanists with the most elaborate equipment available to American science. Now, there is no doubt that we can communicate with plants; at least we know they understand us. We have only to discover or perhaps to invent a common language to learn from them. Antiquity foresaw this possibility and modern science confirms it. It is but a matter of time. Plants are patient. They have all eternity in which to deliver their message.

Calcutta in the days of Jacquemont. Right, entrance to the botanical garden there. (London, Victoria and Albert Museum. Photo by ERL)

APPENDIX

INDEX OF PLANT NAMES

Lille

Varangeville-sur-mer
* Les Moutiers

Cherbourg
La Fauconnière *

* Rouen

* Caen Harcourt
Eltre
* Paris Muséum
* Paris-Pharmacie
* Paris École du Breuil
Bois de Vincennes

Nancy
* Col de
Saverne

Boulogne Billancourt
Jardin Kahn *
Versailles
Chèvreloup et Trianon

* Verrières

École
Amance
La Sivrite
* Jardin botanique

* Strasbourg

Rennes
* Université
* Ville
Le Thabor

Angers
* Pharmacie
Maulèvrie
Allard

Montoire
* La Fosse

Nogent-sur-Vernisson
Les Barres

* Nantes

* Tours

Vernou-en-Sologne
Les Augeres

* Dijon

Besançon
(Le) Haut-Chitelet
Vosges

Villeneuve sur Allier
Balaine

* Chalon-sur-Saône

* Poitiers

Roanne
Grands Muicins

Dompierre
les Ormesi
Pézanin

Samoens
La Jaysinia *

* Limoges
La Jonchère

* Royat

Lyon *

Chateauneuf
sur-Loire

Pharmacie
* Parc de la Tête d'Or

* Bordeaux

* Col du
Lautaret

Hort de Dieu
Généralités
Lafoux

Générargues
Pratrance

Aigoual

* Val Rahmey
Menton
* La Léonie
Beaulieu-sur-mer
Les Cèdres St-Jean C Ferrat

* Montpellier

Antibes
Gratteloup
* Villa
Thuret

* Toulouse

Marseille
* Parc Borély

Font Romeu

Botanical Gardens and Arboretums

of France

This list of selected botanical gardens and arboretums was compiled from the *Jardins botaniques et arboretums de France*, by Roger de Vilmorin (Paris: La Documentation française, 1974). Roger de Vilmorin is a member of the Académie d'agriculture and an honorary research professor at the French National Center for Scientific Research (C.N.R.S.). He was born and raised on the family property of Verrières-le-Buisson, which included a famous arboretum now owned by the city.

▲ ARBORETUM   ○ PRIVATELY OWNED

* BOTANICAL GARDEN   ● PUBLICLY OWNED

An *arboretum* contains only ligneous species, from small shrubs to tall trees.

A *botanical garden* displays either herbaceous and small ligneous species only or species of all sorts, from the herbaceous to trees.

When a garden or arboretum listed below has no specific name, it is called locally the city's garden or arboretum.

APPENDIX

PUBLIC BOTANICAL GARDENS

Antibes Le Jardin Thuret, founded in 1817. Many exotic species acclimatized outdoors. Open every day; guided tour on request.

Besançon Created in 1880. A collection of many ancient gardens, one dating back to 1580, it specializes in medicinal plants. Remarkable greenhouses with subtropical species and small arboretum. Open every day.

Bordeaux Founded in 1726. Many rare plants, a collection of herbaria. The library houses 6,000 books, which can be borrowed. Open every day.

Boulogne-Billancourt Jardin Albert Kahn. Intriguing Japanese garden. Open March 15 to October 15.

Caen Jardin botanique de la ville et de l'université. Created in 1689, damaged during the French Revolution and by bombing in 1944, now completely restored. Superb greenhouses for South American species. Open every day. Guided group tours on request.

Châlon-sur Saône A garden in the making, constantly enriched. Open every day.

Col du Lautaret (Hautes-Alpes) Created in 1894, at 2,100 meters (over 6,000 feet high) — All the rare alpine and snow species. A mountain cabin has been transformed into a laboratory for scholars. Open whenever not blocked by snow.

Col de Saverne (Vosges) A mountain garden, created in 1931. Plants which adapt to sharp slopes, ravines, and rock gardens. Open May to September.

Dijon Jardin de l'Arquebuse. Founded in 1771, and first devoted to medicinal plants. Local and exotic flora, small arboretum, and a botanical school. Open every day.

Le Haut-Chitelet (Vosges)   High altitude garden. Ten thousand species of mountain flora from the whole world. Open June to October.

Lille   Founded in the sixteenth century, and reorganized in an exemplary way. Model greenhouses and ecological classification of plants. Also provides a system of plant reference cards in Braille. Permanently open.

Limoges   Jardin botanique de l'Evêché. A nineteenth century park with exotic trees, and a garden specialized in plants from every European mountain range. A collection of grape vines unique in Europe, and greenhouses among the richest in the world. Open every day.

Marseilles   Jardin du parc Borely. A botanical garden linked to a park, privately created during the eighteenth century. Open every day except Saturday, Sunday, and public holidays.

Menton   Jardin du Val "Rahmeh." Created in 1905 by Lord Radcliffe, ex-governor of Malta. Exotic flora. Open every day except Tuesday.

Monte Carlo   Exotic garden with the most remarkable European collection of cacti and succulents.

Montpellier   Le Jardin des plantes. The most ancient botanical garden in France. Open every day.

Nancy   Created in 1786, and constantly remodelled. Empress Josephine visited it, and offered a collection of the most beautiful plants grown in her garden at the Malmaison. Divided into agricultural and scientific sections. A part of it is permanently open; others are open on request.

APPENDIX

Nantes Founded in the seventeenth century as the ideal relay for exotic plants which had to be acclimatized in France. The first magnolia came from America to this garden. All the rarest species still thrive. Handsome arboretum. Open every day.

Paris Bois de Vincennes—Ecole du Breuil. Botanical and horticultural garden, rose garden, vegetable garden, and orchard. Open only to researchers and students. Organized groups are allowed to visit on request.

Paris Jardin des plantes—Muséum d'histoire naturelle. Founded in 1635, it is the subject of this book. Displays everything. Permanently open, but check the hours for visiting the greenhouses.

Poitiers La station biologique de Beau-Site. Created in the late 19th century, it belongs to the University of Poitiers. Extremely rare species and botanical curiosities. Open only to researchers but guided tours on request.

Rennes Jardin botanique de la ville ou Jardin Thabor founded in 1610. Originally a monks' convent garden. Plants from all over the world. Open every day.

Rouen Founded in 1735. Vast collection of exotic species, especially rare aquatic plants. Open every day, with courses for amateur botanists and guided tours.

Samoens La Jaysinia. Plants and rare species from all the European mountains. Includes a laboratory. Open every day.

Strasbourg Jardin de l'Institut de botanique. Founded in 1619, modified throughout the centuries. Greenhouses with a large display of tropical species. Grouped tours on request.

Toulouse Jardin de l'université Paul-Sabatier. Open only to researchers, but the neighboring public garden has a large collection of trees.

Appendix

Tours   Created in 1840. Important collection. Greenhouses harboring 600,000 decorative plants. Open every day except Saturday, Sunday, and public holidays.

PRIVATE BOTANICAL GARDENS

Beaulieu-sur-Mer   Jardin botanique "La Leonida." Owned by the Plesch family. Made by a connoisseur who gathered himself, throughout the world, the plants he dreamt of. 3,5000 species and varieties on display, internationally famous. Visit upon written request.

Cherbourg   La Fauconnière. A collection of 6,000 species created and owned by the Favier family. Grouped tour on request.

Saint-Jean-Cap-Ferrat   Les Cèdres. Owned today by M. Julien Marnier-Lapostole, it was the garden of King Leopold II of Belgium. The most spectacular garden of the Riviera, with water works, and tropical species. Usually open one day a week. Check for dates and hours.

Varengeville-sur-Mer   Parc des Moutiers. Owned by the Mallet family. "Vegetal scenes" along a predisposed itinerary. This famous garden is open every day.

IMPORTANT ARBORETUMS

Angers   Arboretum Allard ("La Maulevrie"). Exceptional collections of rare species. Open to the public, although most of the delicate plants are contained in closed areas.

Chateauneuf-sur-Loire   Arboretum de la ville. Splendid exotic trees, and a collection of magnolias. Open every day.

Domprierre-les-Ormes (Saône et Loire) Arboretum de Pezanin. Remarkable collection of pine trees from all over the world. Open every day.

Font-Romeu Arboretum de l'Etat. Experimental station for climatic research. Closed to the public.

Gratteloup (Var) Arboretum de l'Etat. Another government experimental station and closed to the public.

Harcourt (Eure) Arboretum de l'Académie d'agriculture de France. Historical arboretum planted around the Harcourt castle and its own handsome park. Michaux's son, André François, planted here the most handsome American tree species starting in 1833. Castle, park, and arboretum open to the public.

Limoges Arboretum de La Jonchère. Devoted to rare trees from the west coast of the United States. Open only to specialists in forestry.

Massif de L'Aigoual (Cévennes) Specialized arboretums. Visitors must obtain permission to visit from the office of the National des Forêts, in Montpellier.

Nancy Arboretum d'Amance; parc de l'Ecole Forestière, and Arboretum de la Sivrite). Rare species. Not open to public.

Nogent-sur-Vernisson (Loiret) Arboretum des Barres. Created in 1821 by Philippe Andre de Vilmorin. Handsome trees from every continent. Open only to researchers.

Roanne Arboretum des Grands Murcins. Many acclimatized species and ornamental shrubs. Open every day.

Royat Arboretum de l'Etat. Forestry station with exotic and American species. Center for ecological research. Visitors must obtain permission from the Office des Forêts in Clermont-Ferrand.

Versailles Arboretum de Chèvreloup. A branch of the Muséum d'histoire naturelle in Paris. André Thouin started his acclimatization here. Species from all over the world. Grouped tours on request.

Versailles Arboretum du Petit Trianon. The trees that Louis XV and Marie-Antoinette loved. Open.

PRIVATE ARBORETUMS

Générargues (Gard) Prafrance. Owned by the Maurice Nègre family. Exotic trees and a fascinating bamboo forest. Open all year round.

Villeneuve-sur-Allier Arboretum de Balaine. Founded by the daughter of Michel Adanson, it still belongs to the family. Classified by the Beaux-Arts. Open permanently.

Index of Plant Names

213